THE EXTRAORDINARY STORY OF HUMAN ORIGINS

THE
EXTRAORDINARY
STORY OF
HUMAN ORIGINS

PIERO AND ALBERTO ANGELA

Translated from the Italian
by Gabriele Tonne

Illustrations by Valter Fogato

Prometheus Books • Buffalo, New York

Published 1993 by Prometheus Books

97 96 95 94   5 4 3 2

Library of Congress Cataloging-in-Publication Data

Angela, Piero, 1928–
    [La straordinaria storia dell'uomo. English.]
    The extraordinary story of human origins / Piero and Alberto
Angela : translated from the Italian by Gabriele Tonne :
illustrations by Valter Fogato.
        p.   cm.
    ISBN 0-87975-803-1
    1. Man—Origin. 2. Human evolution. I. Angela, Alberto, 1962–
II. Title.
GN281.A6313   1993
573.2—dc20   93-9492
                                                       CIP

Printed in the United States of America on acid-free paper.

Contents

6　Contents

Introduction

How far back can you trace your family tree? Most people cannot go beyond their great-great-grandparents. Not even church registers can take us farther back than the seventeenth century (when registers were introduced by the Council of Trent). Before that time, only rare notary documents can shed some light on our past.

True, some aristocratic families can trace their genealogy a little farther back, but not much. Why not? They usually conserve documents of only a small number of their ancestors—the total number, if we were to go back to the Crusades (counting three generations per century and, of course, two parents for each ancestor) would be two billion! Naturally, many of these would be the same people (because the families intermarried continually), but that just shows how difficult it is to reassemble the mosaic of our ancestry.

Even if we were able to go back as far as the Roman and Etruscan periods, all traces would be lost after that: no written documentation was kept before that time, even in the most noble families. And the Romans lived only twenty centuries ago; our most distant ancestors lived forty thousand centuries ago. Accounting for twenty centuries is like crossing Manhattan, but accounting for forty thousand is like going around the world.

And yet, "records," "documents," and "evidence" of these forty thousand centuries exists, which allow us to write (although in quite a different way) the history of our ancestors. This "evidence" is the sediments, footprints, and fossils that petrified in the folds of the earth over the course of the centuries and that have only in recent decades been the subject of systematic study by scientists.

It is not easy to unearth knowledge in this field, but it is extremely exciting. One of the most fascinating enigmas on earth is now being investigated—the origins of the human species. Like so many Sherlock Holmeses, researchers seek all possible clues and analyze them meticulously in hopes of unraveling the mystery.

That is exactly what we will attempt to do in this book. Like a detective story, it will try to fit together the pieces of the puzzle available today. Our adventure into prehistory will be aided by experts in paleoanthropology who have studied these clues and have tried to extract as much information as possible through paleontological, anatomical, biochemical, geologic, genetic, and physical analysis. Just as a cigarette butt, a hair, a button can provide the key to identifying the "culprit," the layers of a fossil, the way in which a rock is chipped, or the detail of a joint can provide important information on the life, the appearance, and the behavior of hominids. The most distinctive feature of the study of the past is that it has to make do with the few, often extremely fragmented, remains that researchers have been lucky enough to come upon. This demands great attention to detail.

We would like to thank all those whose interviews helped us illuminate the past: Luca Cavalli Sforza (Stanford University); Yves Coppens (Collège de France); Giacomo Giacobini (University of Turin); Jan Jelinek (Moravske Museum); Donald Johanson (Institute of Human Origins, Berkeley, California); Richard Leakey (National Museum of Kenya); Henri and Marie-Antoinette de Lumley (Institut de Paléontologie Humaine, Paris); Marylène Patou (Institut de Paléontologie Humaine, Paris); Carlo Peretto (University of Ferrara); Alberto Piazza (University of Turin, Department of Genetics); Marcello Piperno (Museo Pigorini, Rome); Pierre François Puech (Nimes, France); Brigitte Senut (Musée d'Histoire Naturelle, Paris); Christopher Stringer (British Museum, London); Phillip Tobias (University of the Witwatersrand, Johannesburg), Erik Trinkaus (University of New Mexico at Albuquerque); Alan Walker (Johns Hopkins University, Baltimore); Tim White (University of California, Berkeley); and Bernard Vandermeersch (University of Bordeaux).

We would also like to express our gratitude to all those whose work enabled this investigation into the origins of the human species to provide answers to some basic of questions. First of all, to Louis and Mary Leakey, who as early as the 1930s made fundamental contributions to this study; and then to Clark Howell, Sherwood Washburn, David Pilbeam, Owen Lovejoy, Desmond Clark, André Leroi-Gourhan, and all the others whose work will be referred to in this book.

A special thanks to the Centro Studi e Ricerche Ligabue in Venice and its untiring director, Giancarlo Ligabue, for the opportunity to take part in his research projects and expeditions, some of which dealt with the very subject of this book.

For our part, we will make every effort to be as clear and comprehensible as possible, offering the reader a detailed (but hopefully not boring) guide for this trip back to the origins of human beings.

Our journey begins on a cloudy day on the African savannah, about three million, seven hundred thousand years ago.

1

Looking for Evidence

One Day, 3.7 Million Years Ago

Like all good detective stories, ours starts with footprints in the mud. . . .

The year: 1978. An early summer morning. Huddled around a few square meters of ground, researchers are carefully removing a crust of soil from one of the most extraordinary sights for a paleontologist: a series of footprints left in volcanic ash over three and a half million years ago by two or perhaps three hominids. We are in Laetoli, a locality to the south of the Serengeti Park in Tanzania. From here, our ancestors, walking erect as we do, made their way northward across the savannah.

Where did they come from? And where were they going?

The findings at Laetoli provide a good beginning for our story. The footprints trace a fine line between two worlds: beyond them is the human race, behind them the shadows of evolution. Almost nothing remains of the history of our species prior to these footprints; a long trail of findings providing the basis for theories and debate follows.

How can footprints, which are so easily erased by wind, rain, vegetation, or other footprints, be preserved for 3.7 million years? The footprints left by the astronauts on the moon may well be the only others to last for such a long time (given the absence of water, atmosphere, and life on the moon).

As in Pompei, a unique event took place at Laetoli: volcanic ash covered the footprints and preserved them. Meticulous study of the ground has revealed that the hominids were walking on a layer of ash deposited by an erupting volcano and that the ash was impregnated

11

1. Footprints in the sand at Laetoli: thorough examination has revealed that they were left on a rainy day 3.7 million years ago by beings that walked in an upright position like us.

with rain water. It was the end of the dry season and hyenas, gazelles, elephants, and rhinoceroses roamed the area.

But how can all this information be gained from the soil?

Clues and Evidence

To start with, the volcano: Lapilli (i.e., small rocks) found in the area originated from the Sadiman volcano located to the east. But the ash would not have been able to settle on the savannah if there had been tall grass; thus, the grass must have been eaten by animals (their footprints have been found), and therefore it must have been the dry season. But the astonishing thing is that even traces of raindrops have been found. The first rainstorms were brewing; it was probably the beginning of the rainy season.

This unreal setting was the backdrop to the northbound exodus of two or three hominids. Strangely enough, they seem to have crossed that stretch of savannah a number of hours (or days) after the eruption of the volcano; other fine layers of volcanic ash from the same eruption were found under their footprints. Thus, they had already been "on the road" for some time. Did they come from far away or from the vicinity of the volcano? We will never know.

What is certain is that they were not running: the distance between their steps clearly proves that they were walking. They may, however, have been frightened by the events around them: the layer of ash they were walking on was already ten centimeters thick (that is the thickness of the crust under the footprints), and the entire savannah must heve been transformed by this volcanic "snowfall," with its smoke, lapilli, and fires. Molded by the particular combination of ash and rain water, the footprints were immediately covered by more volcanic ash, which sealed them and preserved them like photographs of a distant past.

But why are these footprints so important? Because they reveal that at that time, over three and a half million years ago, bipeds existed. Not only that, their footprints are so similar to ours that they are almost indistinguishable.

This will be discussed in the next chapter, in an attempt to understand what these hominids looked like. But Laetoli has already provided an example of the kind of information that can be deduced from paleontological remains. In order to get this information, of course, the remains must first be found. And that's the difficult part. How does one go about it?

Fossilization: How Bone Turns to Rock

All living beings, whether human beings, animals, or plants, decay rapidly after death, especially if left out in the open. With a few exceptions (such as animals trapped in ice, like the famous mammoths in Siberia, or insects trapped in resin which then turns into amber), usually nothing remains. The body disintegrates: first the soft tissues and gradually all the rest.

The skeleton takes the longest to disintegrate, in particular, the parts made of more compact bone, such as the jaw, the skull, and, of course, the teeth. But they, too, slowly decay; the time required depends on the surroundings. Let's take a gazelle as an example to understand this crucial process.

Immediately after death, the gazelle is torn apart by hyenas and jackals, which scatter its flesh and bones. The bones, already crushed and gnawed, are then attacked by bacteria and atmospheric agents—sun, wind, rain, drought, humidity, and temperature changes—until they are broken down into tiny pieces. The trampling of other animals pulverizes the bones before they are washed into the soil. No trace remains.

And yet, in some cases organic structures have been preserved almost intact to the present day. This happens when the (extremely rare) phenomenon of fossilization takes place. The remains are quickly enveloped in a protective environment. The most typical example (without going into carbonification, incorporation into mineral wax, natural mummification, etc.) is rapid covering by a fine layer of sedimentation. This occurs, for example, when an animal dies on the shores of a lake or the banks of a river, or when it is buried by a landslide or a flood. In this case, the sediment seals the remains, insulating them from destructive agents. Only seeping water reaches the remains and gradually replaces the organic molecules with the molecules of its dissolved mineral salts.

The process could be compared to a bus with people getting on and off: at the end of the line, none of the passengers are the same as those at the beginning, but the bus always looks the same (this also happens in our metabolism). Fossilization takes different lengths of time, depending on the environmental conditions. Usually it takes many thousands (sometimes tens of thousands) of years. An incomplete process produces so-called "sub-fossils."

The kind of minerals found in the soil determine the color and the texture of the fossils; they can be grey, whitish, brown, or even

2. Two hominids who died in different circumstances. The body of the first, who died in the open, is soon picked apart by predators and disintegrated by the elements. The body of the second sinks into the mud of the lake bottom and decomposes, but the bones slowly fossilize, perhaps to be discovered millions of years later.

black. Because of the gradual substitution of the molecules, the fossils look exactly like the originals; their structures appear the same even under the electron microscope.

It is clear that fossilization is a rare event which takes place only under very special circumstances. That is why it is so difficult to find fossils. As Professor Phillip Tobias once commented, "Hominids usually died in the wrong places for fossilization."

To make matters worse, fossils are by definition hidden. It takes only a few centimeters of soil to make them invisible to the human eye (they cannot be detected by radar or other sensors because they are simply minerals, like any rock).

How, then, can we find fossils?

Where to Look for Hominids

The first step is to look for exposure of the desired strata (with the help of geological maps) and hope that erosion will have uncovered something of interest.

In order to do so, though, the right stratum has to be identified. In a search for dinosaurs, the exposures have to be 60, 100, 200 million years old. In a search for hominids, the exposures can be 4 to 6 million years old or less.

These ancient strata are brought to the surface not only by erosion, but also by the slow tectonic movements of the earth's crust. They produce an undulatory effect similar to that created when you push a carpet with your foot. That is why marine fossils can be found on mountain peaks (although this was once used as "proof" of the biblical Flood).

Certain strata fold completely and split open, causing even deeper sediments to emerge beside the more recent ones. In certain places (particularly visible in desert areas) ground which is only 5 million years old can be found between ground which is 80 million years old and other patches which are 30 million years old. Thus, within only a few miles, we can move from the Cretaceous period to the Pliocene to the Oligocene, with the same ease as if passing through different municipalities on a highway.

But in order to find hominds, we have to look not only in the right stratum, but also in the right region. The hominids gradually expanded into certain regions and not others.

It would be of little use to look for *Homo habilis* in Sweden, for example, even if in the right stratum. It is believed that human beings left Africa only much later, probably as *Homo erectus*. Similarly, there is no guarantee that looking in the right area (and the right stratum) will lead to discovery of the right fossils.

Where to look, then? A paleontologist once said jokingly, but with a trace of seriousness, that the best way to find fossils is to go where others have already found them. By this he meant that since these places are obviously suitable for fossilization, and since erosion will continue to reveal new strata, each year may reveal something new which was only a few inches beneath the surface the year before.

At the moment, south and east Africa (particularly Ethiopia, Kenya, and Tanzania) are the areas in which the most ancient hominids have been found. These areas continue to provide the most important finds year after year—sometimes in places researchers have been going to for years already. That is what happened in 1986 in Olduvai (Tanzania) when the team of Donald Johanson found *Homo habilis*, the so-called OH62, only a few feet from a busy road.

There are probably many other places in Africa suitable for this kind of quest. If we look at the map, in fact, it seems absurd that the remains of hominids can be located only in south and east Africa and not in the interjacent areas. Let's ask Professor Phillip Tobias of the Witwatersrand University in Johannesburg, South Africa:

Why this gap, Professor Tobias?

Surveying has been carried out in various regions, but the problem is that there are no exposures from the right period.

Some excellent sediments are located in the north of Malawi, but no hominids have been found. But one must not forget that it takes a lot of time and tenacity. The Leakeys were in Olduvai for thirty years before they came upon a hominid in 1959. They knew that they had to be there because they had found stone tools and their determination paid off. Since then the remains of many hominids have been found in that area—sixty-four in all.

So a lot still has to be done.

Oh yes. We know that there are probably thousands of hominids just waiting to be unearthed.

Why have all the fossils been found in eastern Africa? The western part of Africa is covered by vast forests; has that made excavation a problem, or is it simply felt that hominids could evolve only on the savannah and not in the forest?

Let's ask one of the greatest fossil-hunters alive today: Richard Leakey, son of Louis and Mary Leakey, and former director of the Museum of Natural History in Nairobi, Kenya:

It's hard to say. I believe that since tropical forest covers most of western Africa today, it probably has been that way for millions of years. Even during the evolution of the human species.

Therefore, if we are the result of adaptation to an open environment, we shouldn't expect to find fossils from primitive stages in western Africa and don't even have to bother to look.

Some surveys have been carried out in Cameroun, northeastern Nigeria, and other parts of western Africa, but nothing has been found.

In Search of the Site

But once in the right place and with the right strata, what does a good investigator do?

An extraordinary thing happened to us while on an expedition to the Egyptian desert: when we woke up one morning, we found ourselves in a prehistoric camp. Stone tools and chips were scattered all around, and we even found the remains of ostrich eggs (which became extinct in that area thousands of years ago) and a campfire.

Does that mean that human beings instinctively tend to camp in certain places, for geographic and strategic reasons? This may be the case for more recent settlements (thousands of years old) but it can hardly be true for settlements hundreds of thousands (or even millions) of years old; the environment has changed too much since then.

So, the thing to do is to try to imagine what the former environment was like and identify the most suitable places for human settlement, for example, along the banks of lakes and rivers that formerly existed. Not only is water essential for human life, the probability of finding animals in its vicinity (either to hunt or take away from predators) has made it fundamental throughout human history. Moreover, it creates a suitable environment for fossilization.

How can these sites be found once everything has changed? This is where experts and their laboratory techniques come in. Soil analysis can provide detailed information about the former nature of the area: for example, it can distinguish the bottom of what was once a lake from the shore (and can even calculate the intensity of the waves on the shore).

In the future this technique may be assisted by satellite observation. Radar surveys of the Egyptian Sahara have revealed a network of rivers which have long since dried up and disappeared. Today, they are invisible from the ground because they have been covered over.

These new forms of investigation have enabled archeologists to excavate in particularly favorable places, such as ancient river banks. The result has been the discovery of stone tools dating back hundreds of thousands of years. Some scientists have hypothesized that these rivers helped the hominids migrate toward the north. Perhaps *Homo erectus* passed this way on his gradual move toward Europe and Asia about one million years ago.

In later chapters be will see how other techniques can be used to "read" the terrain and sites for more information. But the most indispensable, irreplaceable, and refined instrument of all in human paleontology—the product of millions of years of evolution—is our eyes.

Our Eyes: Our Most Precious Instrument

Direct on-site observation is still the best research method. A researcher can have expertise, technology, and scientific knowledge; but without the ability to scan the ground, a stroke of luck is more unlikely.

The remains of hominids do not come to the surface intact, with the skull and the skeleton carefully ordered. Most of the time only fragments or broken pieces are found. Even skulls are rarely intact. Especially in Africa, finds are usually just chips, no larger than a stamp, almost indistinguishable in size and color from the surrounding pebbles.

It takes a researcher's trained eye to spot them. Close examination of the surrounding area and excavation often reveal other fragments that must be pieced together patiently later.

Let's find out what Donald Johanson of the Institute of Human Origins in Berkeley, California, who discovered "Lucy" and many other hominids (Australopithecines and *Homo habilis*) has to say:

I have always come upon my best finds while concentrating intensely on what I was doing. It's easy to be distracted while out in the field. One's mind starts to wander, perhaps to one's problems. I remember that when I found the fossils in Hadar, I has thinking of that alone, nothing else.

Like an athlete concentrating before a race?

Exactly. Field work calls for total concentration. The mental state is all-important. You have to focus on tiny things; boulders, hills, slopes have to disappear.

It's almost second nature. I picked up the habit as a child when an anthropologist used to take me out for walks. He would suddenly ask me what I had seen in the last ten minutes. If I had missed something, we would go back.

It's become a habit. No detail escapes me, even in a room or a store.

Being a good observer is an essential quality which can be improved. Many young researchers who start work in the field think that fossils can be found quickly and easily. When they don't find them, they are disappointed.

Stone Tools: Man Is Near

One of the most important signs of the presence of man is stone tools. It's like finding empty cartridge shells in a forest: you can be sure that hunters have been there. Extreme caution must be exercised in assessing the kind of tools and the way they were made (more on this later); this can lead to an understanding of the period in which they were manufactured. Discovery in strata from a certain period (*in situ*, as they say) provides proof of their authenticity.

At that point, it's exciting to look for fossil remains of the hominids who used them. They are bound to be there somewhere, it's only a matter of finding them.

As in all human matters, luck or chance plays a role. But talent makes the difference, especially when surveying a gorge or a depression, or choosing the right time and the right light. It's something like looking for mushrooms. Some African specialists, such as the Kenyans Kamoya Kimeu and Kiptalam Chepboi, are extraordinary prospectors, and have made some great discoveries possible.

Often, researchers think they have hit on something, start to get excited, and call their colleagues in—only to find that the fragments of bone statistically belong to various animal species. Disappointment is part of the process.

But so is excitement. Sometimes—actually quite rarely—the remains of hominids are found. Only then does the real investigation begin: Was it a male or a female? What was this creature like? How did it live? What was the environment like? What did it eat? How old was it? Did it have any illnesses? Above all, what link does this creature represent in the chain of human evolution?

The Excavation: Like a Crossword Puzzle

When researchers come upon the remains of a hominid, they act rather like police investigators who discover a corpse: nothing can be touched, no object moved. Even the most minute detail could turn out to be essential for the reconstruction of the hominid's identity.

The site is fenced off and divided into one meter squares by strings. Like a crossword puzzle, each square is numbered and excavated by different members of the expedition. Nothing must be lost, so very fine instruments and brushes, similar to those used by a dentist, are used in excavation. Before being removed, the position of each fragment is sketched onto a map indicating orientation and angulation.

Consolidating substances (or plaster casts) are used to prevent the fragments from crumbling. In this way, they can be removed from the site without damage and later freed of their support in the laboratory. Each stage of the operation is carefully photographed.

During excavation, not only the remains of the hominid but everything around them is important. Just as a word gains significance in relation to the other words in the sentence, so must a hominid be seen in context. As a result, the soil, the microfossils, the animal remains around it must be studied. Thanks to modern technology and scientific techniques, all of these "silent witnesses" can provide an enormous amount of information. Let's find out how.

What Can We Learn from the Soil?

For the average person, the soil is the same everywhere; it is of no importance. It used to be shoveled away in the early paleontological excavations. Today we know that it can tell us a lot of things.

The soil is carefully put through different-sized screens to sift out various kinds of remains and is then examined in the field and in the lab. For example, microscopic analysis of the forms of the grains of sand can reveal whether the sand comes from the desert (aeolic origin) or from the banks of rivers, lakes, or seas; the grains from the desert are rounded and levigated from the constant rolling action of the wind, whereas the grains originating on river banks are more angular. Similarly, other kinds of sediments (clay, marl, silt, or gravel) provide paleontological detectives with other information on the environment—the presence of lakes, rivers, deserts, and so on—and the climate—hot, cold, humid, or otherwise.

The soil can also provide another invaluable source of information: pollen and spores. These are direct evidence of (at least some types) of vegetation that existed. Microscopic examination of these tiny plant fossils (their sizes range from 5 to 200 thousandths of a millimeter) provides new climatic and environmental data. This kind of examination can sometimes lead to extremely interesting discoveries: for example, that flowers were placed on the burial mound of a Neanderthal in Iraq.

But the soil can also provide investigators of the past with other information, such as the presence of fire or pieces of wood used for the fire. In this way, the kinds of trees and shrubs that grew in the area can be identified, giving indirect information about the climate. The time at which the combustion took place can be dated: analysis of the carbon 14 contained in bits of charred wood allows for dating as far back as 35,000 years (and with an accelerator, even 70,000 to 80,000 years).

The discovery of tiny fossils of so-called micromammals, such as rodents or insectivores, in the same stratum can also help in the dating of a site. The evolution of these animals was so rapid in certain periods (thanks to their high rate of reproduction) that a given species can provide a specific date. They are almost like fossil calendars. Conveniently, the most precise period of the calendar coincides with the period in which much of the evolution of the human race took place.

The importance of other kinds of finds, in particular the remains of birds and mammals, will be discussed later. But it is already quite clear that an enormous amount of information can be obtained from the soil and from the other things found in situ (especially stone tools).

Yves Coppens, professor at the Collège de France, former curator of the Musée de l'Homme in Paris, and leader of numerous excavations all over the world, predicts that other techniques will probably be discovered in the future:

I'm sure that research techniques providing us with a much more complete overall picture for each period will be developed in the coming years.

And we'll regret the way we excavated in the past, or perhaps even the way be excavate now?

Yes, in the future we will probably regret our current method of excavation. For example, when I first started, each time a tool was found, it was brushed. . . . Today it is observed under the microscope, magnified ten to twenty thousand times, for the most minute traces. Examination of the blade of a knife, for example, can reveal whether it was used to scrape a bone or to cut wood, a vegetable, or tendons.

The Piltdown Case: A Paleontological Fake

Just as lip marks on a glass can lead a detective to the person who drank from it, so can everything that surrounds the remains of a hominid provide useful information about that creature.

A detailed explanation of how all elements can contribute to the dating of a find is given in Appendix 1. This fascinating subject cannot be dealt with here, but the interested reader may turn to the back of the book and read about these techniques right away.

Thus, chemistry, botany, physics, biology, climatology, and genetics have all become involved in human paleontology, making it a multidisciplinary science. The interrelations that exist today safeguard it from hoaxes like the famous skull found at Piltdown.

Many of you will remember the Piltdown case. In 1912, in an English locality of the same name, an extraordinary skull was found which was taken by some to be the missing link between humans and apes. A partial reconstruction of the pieces presented a human skull with an ape-like jaw. Only forty years later, in 1953, was it revealed to be a well-prepared fake. The medieval cranium and the orangutan mandible were so well matched that they seemed to belong to the same skull (they had been found together).

The perpetrator of the hoax had painted the two pieces with a substance to give them the same ancient patina and had filed some of the teeth to create a human bite. The mandible had astutely been broken to match. Fossil fragments of animals of the past, as well as some flints and some bone tools, were scattered in the area.

The English paleontologists of the time were completely taken in by hoax—they did not have the techniques needed to unmask it.

The Piltdown case became the subject of controversy for a long time: many found that the reconstruction lacked credibility and that pieces of human bone had been associated with pieces of animal bone. Some openly suggested it was a fake. But the hoax was only revealed in 1953, when (after analysis with fluorine had raised further suspicion) a piece of the mandible was perforated and it was discovered that it had been artificially colored.

Could something like this happen today?

"Absolutely not," says Yves Coppens.

Too many laboratory and field tests for authenticity exist today. I've got a short anecdote about that. During an expedition to the Omo Valley, one of our assistants had such a terrible toothache that we had to pull the tooth. I decided to play a trick on our American colleagues and secretly hid the tooth in the strata they were excavating with Clark Howell! More than ten years have gone by since then and nothing has ever been published about the discovery of that tooth. They must have identified it for what it was.

What Has Sherlock Holmes Got to Do with It?

Some mystery still revolves around the Piltdown affair: whose idea was the hoax?

The most obvious suspect was Charles Dawson, the amateur geologist who signaled the find. (He was a solicitor whose passion for fossil hunting provided the British Museum with some fine specimens.) He claimed he had been attracted to the spot by previous findings of flints.

In 1915, Dawson came upon a second find, four kilometers away (fragments of another skull, also later found to be a fake). Dawson died shortly afterward, without a word of explanation. Was he the perpetrator of the hoax? Was it meant as a joke? Why did he do it? To establish a reputation for himself? We will never know.

But Dawson was not the only suspect. Some evidence points to Sir Arthur Conan Doyle, the creator of Sherlock Holmes. If he had been involved in the affair, it would have underlined his skill not only in coming up with evidence to solve a case, but in inventing cases to be solved using evidence. Unfortunately, no one was able to solve this case for many years.

Before ending this chapter on the search for hominids (and the various techniques used to do so), some very special findings must be mentioned.

In research, as in life, very unexpected and strange things sometimes happen. Important discoveries do not always occur in the traditional way. Sometimes, as we will see, they take place by chance, in the most unusual, even paradoxical, ways. Of the long list of important chance findings, only four will be mentioned: Peking Man, Saccopastore Man, the thirteen hominids of Hadar, and the footprints at Laetoli. Each of these has a very special and instructive story to tell.

An Unexpected Deposit: The Druggist's Shop

Let's start with Peking Man. This is an example of how local tales, oral tradition, and even fruit and vegetable markets can give rise to important discoveries.

At the end of the last century, while wandering through some Chinese shops, a German naturalist, K. A. Haberer, noticed that the "dragon's bones and teeth" sold by many druggists were actually the fossil remains of ancient mammals. Pulverized, they were mixed with "magic" ingredients and sold as miracle potions.

He realized that there must have been a rich deposit in the area. Word soon spread, and in the years that followed a number of European paleontologists were drawn to the region. One of them, an Austrian named Zdansky, found some hominid teeth in Zhoukoudian. They belonged to a *Homo erectus* several hundreds of thousands of years old. However, he kept the discovery to himself for five years, since he was afraid that the sensation would ruin his work, which was actually focused on the fauna in the area. The year after his announcement, Davidson Black, a Canadian anatomist teaching in Peking, found another tooth in the same area, which he claimed belonged to Peking Man.

Research carried out in Zhoukoudian between 1929 and 1937 led to the discovery of the remains of over forty individuals: an extraordinary collection which came to an extraordinary end. In 1941, after the Japanese invasion of China, the fossils were packed into two large crates for shipment to the United States. Legend has it that the ship ran aground and that the sailors were all taken prisoner by the Japanese. At any rate, the crates were never seen again.

A Swift Blow to the Head

The story of the Saccopastore Man is much simpler. Saccopastore is a locality on the outskirts of Rome, on the left bank of the Aniene River. In spring of 1929, a worker digging in a gravel pit sent his pick through the skull of what turned out to be a pre-Neanderthal who had lived between 120,000 and 130,000 years ago.

Taken by curiosity, the worker tried to remove the strange object that partially projected from the mass of gravel into which it was embedded. During this rather rudimentary operation the brown ridge (the famous receding forehead of prehistoric man) and many teeth were lost.

Fortunately, after a few days in the shack at the pit, the skull was handed over to the owner, who immediately passed it on to a famous scholar in the field, Sergio Sergi. Six years later, during a break in a paleontological conference, Alberto Carlo Blanc and Abbot Henri Breuil returned to the pit which had been abandoned in the meantime and quite accidentally, without even digging, came upon another smaller and less complete skull. This was the skull of a man (the first was of a woman), about five thousand years older.

That was not the only time in the history of paleontology that a find came to light in a violent way, that is, not through the gentle touch of the archeologist's instruments and brush. Another famous find, in Taung in South Africa, occurred when a dynamite explosion in a rock quarry split and exposed what is still considered one of the most important specimens in paleoanthropology: the skull of an *Australopithecus* child, five or six years of age, who lived two million years ago.

Fortunately, a geologist was also present at the scene. He packed up the skull and handed it over to an expert, Raymond Dart.

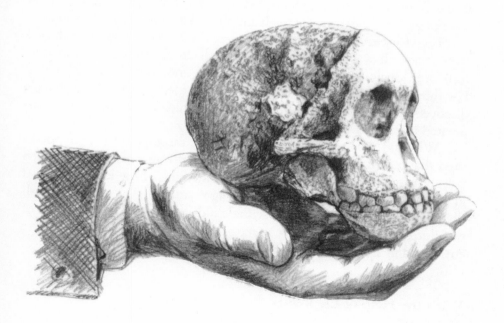

3. Our first image of an Australopithecine from the past: the Taung skull of a five- or six-year-old child. The finding gave new impulse to the investigation into our most distant ancestors.

A Cushion of Hominids

The story of the discovery at Hadar (site 333) is quite different. A group of paleontologists was already at work in an area of Ethiopia known to be rich in fossils. While exploring in autumn of 1975, Donald Johanson and his assistant came upon two premolars of an Australopithecine who had lived at least three million years ago. Two French filmmakers and a photographer from *National Geographic* magazine were guests at the camp at that time. They all decided to return to the site together the next morning when the light would be right for photographing the find.

While they were reconstructing the scene, the cameraman's wife, who had absolutely no knowledge of paleontology, decided to withdraw to the shade of an acacia tree halfway up the hill. When she sat down, she felt something sticking up out of the ground. Picking up the two pieces of bone, she waved and called out, "Hey! What are these?" They were pieces of bone from the thigh and the ankle of an Australopithecine.

Feverish excavation began around the tree. More and more fossils were found. It was an unexpected windfall. In all, the remains of thirteen individuals were found: men, women, and four or five children! For the first time, the discovery—and as it turned out, the largest single discovery ever made—of the remains of hominids was thoroughly documented by photographs and on film.

Why did thirteen individuals die in the same place? It is hypothesized that a sudden flood took them by surprise in a place from which there was no escape, such as a gorge. (Something of the kind happened in Petra, Jordan, about twenty years ago: sudden rain caused a flash flood, overwhelming a class of schoolchildren on tour.) The bodies of the thirteen Australopithecines were then completely submerged and enveloped in sediments, and slowly transformed into fossils.

With the Help of Elephants

To end this chapter, the last discovery to be made in unusual (and in this case rather amusing) circumstances were the footprints in Laetoli. The expedition of paleontologists was led by Mary Leakey.

The quest had been long and tiresome. But there were moments for rest and laughter. During one of these breaks, two younger members of the expedition took up a rather unacademic game—throwing elephant dung at each other. In an attempt to dodge the other's projectiles and to find some raw material himself, one of the two came upon something unusual. The game was immediately suspended and upon closer scrutiny, the unusual object turned out to be a pile of volcanic ash bearing animal tracks. That was the summer of 1976 and the beginning of the search for other footprints in the Laetoli area. Two years later, the now famous footprints—an important milestone in paleontological history—were discovered.

Let's go back to those footprints and continue our story where we left off. Where were the hominids coming from? At what stage on the evolutionary scale were they?

Laetoli is the starting point for our trip back in time in search of evidence that can shed light on the forerunners of human prehistory.

2

The Forebears of Our Ancestors:
The Pre-Hominids

What the Footprints at Laetoli Tell Us

Footprints left in the mud under a window can provide an investigator with quite a bit of information: for example, whether there was one or more persons, whether they were male or female, what size (and what kind of) shoes they were wearing, more or less how tall they were, and possibly how much they weighed (if the pressure exerted on the mud by the body weight can be calculated).

All these analyses were carried out on the Laetoli footprints. And what were the results? Unfortunately, these beings did not wear shoes, so it was impossible to deduce whether they were male or female from their shape. But it is quite evident that the two footprints are different: the ones on the right are bigger and "heavier" and the ones on the left are smaller and "lighter."

The footprints continue for about fifteen feet in two parallel lines no more than about ten inches apart. This is too little for the two individuals to have been walking beside each other.

It could also mean that the hominids left the footprints at different times: in fact, the footprints in the first row (the one on the right) have tiny splash marks all around them, as if the ground were muddier when they were made. Those in the second row (on the left) seem distinct, as if they were made on harder ground.

But some people claim that there is a third row of footprints,

31

underneath the first. This would mean that someone was following earlier footsteps.

Another interesting observation has been made about the left footprints. it seems as though the hominid stopped at a certain point and turned around to look to the left. At what?

Each step measures approximately fifteen inches and the height of the hominids has been calculated to be three and one-half and four feet, respectively (slightly shorter than the Pygmies today).

These footprints have also been examined using a technique called "photogrammetry" (in which the various levels of depression are marked, as in topographical mappings of mountains and valleys). Using this method, the investigators have been able to confirm something that was evident at first glance: the hominids walked in the same way that we do. Weight has borne in the same places: the heel and the tip of the first metatarsal, with the arch raised and the big toe aligned rather than separated.

In other words, the footprint is much like a human footprint, even though there are some differences in the bone structure.

This is the real starting point for our journey into the past: the upright position was perfectly normal 3.7 million years ago. The hallmark of our species—our upright position (no other primate is bipedal, although some other animals are, as will be seen later)—was already completely developed at that time.

The first point this suggests is that evolution toward the erect position must have started much earlier. How long ago?

The Feet and the Brain

This kind of adaptation must have taken a long time because it required a series of correlated transformations, not only of the feet but also of the entire bone structure (from the ankles to the hips, pelvis, and spine), as well as relative changes in the muscles, joints and internal organs, and the circulatory system.

It can be safely assumed that many of the ancestors of the Laetoli hominids were bipeds. That means that the upright position must be an ancient characteristic, even though evidence of the exact date is lacking.

Another starting point is the skull. Although their feet already closely resembled ours, what were their heads (and, therefore, their brains) like?

Data seem to indicate that theirs were quite different from ours. Their skulls must have been much closer to those of the apes, with a much smaller brain volume than we have.

No good fossil remains of skulls have been found at Laetoli (only fragments and teeth), but indirect evidence is provided by the remains of other hominids who lived in that part of Africa. Many skeletons of hominids who lived in a much more recent period (the date is uncertain, but the period is set between 2.9 and 3.7 million years ago) have been found farther north, in Hadar, Ethiopia. One of these is the famous skeleton of Lucy.

The general structure of the body of these hominids, known as *Australopithecus afarensis* (from Afar, where Lucy was found), is quite similar to that of human beings. The teeth are at a stage of development halfway between those of apes and those of human beings. But the brain is still very primitive. It seem as though evolution started at the feet and slowly worked its way up.

If those were the characteristics at the time (that is, a perfectly developed upright position, but still largely underdeveloped brain), two conclusions can be drawn:

1) Contrary to former views, human beings did not "stand up" in order to use tools: the use (and the manufacture) of tools called for a much more developed brain. If the ancestors of such primitive beings already walked on two legs, they must have done so for a different reason (which will be discussed shortly).

2) In order to retrace our genealogy, we have to look for evidence of *bipedalism* and not *intelligence*. This is, however, very difficult, in the absence of sufficiently complete fossils. There is a "black hole" between 5 and 9 million years ago, in which practically no fossils of hominids have been found. Why not?

Because there are no "right" places. The sediments in the places that were found indicate that the environment was not suited to the life and the fossilization of these hominids.

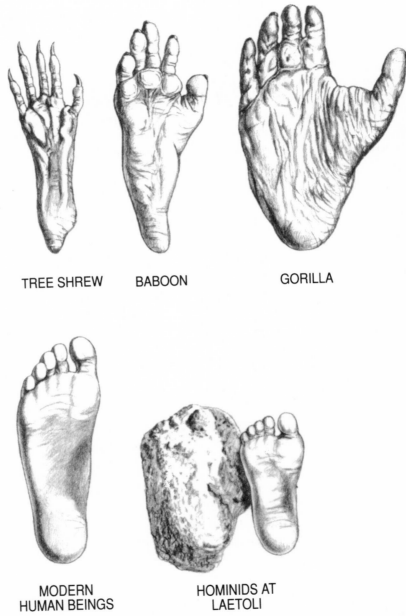

TREE SHREW   BABOON   GORILLA

MODERN
HUMAN BEINGS

HOMINIDS AT
LAETOLI

4. Feet: variations on the theme. *From left to right, top*: the foot of a tree shrew, of a baboon, of a gorilla (not in scale). *Bottom*: The foot of a human being beside the reconstruction of the foot that left the print at Laetoli. The similarity is striking.

But a more general problem is that these strata have not come to the surface. Let's hear that Richard Leakey has to say:

> The layers in which fossils can be found are covered by more recent strata. It's difficult to find erosion that has uncovered the right layers. Some suitable places are located in Ethiopia and probably also in Kenya, but it will take years to discover and study them.
>
> In any case, I think that the reasons for this famous hole are strictly geological.

Of course, the fact that fossils are hidden under the soil, perhaps only at a few centimeters, is a problem. The only instruments you have are your eyes?

Yes. And, of course, one mustn't forget that large parts of Africa are covered with vegetation. There is little erosion and these "right" layers do not emerge.

Have no technologies been developed for underground detection?

No. X-ray shots have been taken from satellites, but they are of more use to archaeologists than to us. And in any case, we'll never have enough money to buy such expensive instruments. Field work is the only solution for now.

Research goes on and will hopefully bear fruit in the future. (Unfortunately, not only are the funds and instruments for this kind of research lacking, but the necessary authorizations can often not be obtained: many sites are, for example, located in areas where guerrilla warfare is going on.)

The Tracks of Evolution

But can't other fossils, such as those of the remains of the great apes that abounded in Africa at the time, be helpful in our investigation? Unfortunately not. Those animals were essentially arboreal, that is, they lived in trees in the tropical forest, and the forest is not conducive to fossilization: the soil is very acid and the combined action of the soil and bacteria erodes the bones before they can fossilize. In fact, there are practically no fossil remains of the ancestors of chimpanzees, gorillas, and orangutans.

Given the "black hole" between 5 and 9 million years ago, the Laetoli footprints (3.7 million years ago) are like a railway station with tracks quite evidently running in our direction (although the route is still rather vague, as we will see). The tracks leading away from the station in the other direction, however, soon fade into the sand.

Only a few fragmentary remains predating the "black hole" have been found: a piece of skull and the upper part of a femur in Ethiopia (Maka and Belohdelie) and a piece of humerus in Kenya (Kanapoi), all around 4 million years old; a piece of mandible with teeth in Lothagam, thought to be about five and a half million years old; and a molar in Lukeino perhaps six and a half million years old. Far too little to be able to piece the puzzle together. A new series of finding made in Ethiopia in the fall of 1992 and the beginning of 1993, consisting of the bone fragments of probably twelve separately found individuals, may offer new insights into the early part of this intriguing period (4–5 million years ago). Study of the specimens is now in progress.

A little farther on, scattered pieces of "rail" and "ties" have been found: for example, a maxilla with a few teeth found in Samburu, Kenya, and studied by Professor H. Ishida. It supposedly dates back 8 or 9 million years and belonged to an ancestor common to human beings, gorillas, and chimpanzees (but not orangutans). This is taken as evidence that the human species had not yet diverged as a specific evolutionary branch at that time, and that this ramification occurred in the period of the "black hole" (between 5 and 8–9 million years ago).

In 1958, another "tie" was found in a mine in Monte Bamboli, Italy, dating back to the same period (8–9 million years ago): a splendid specimen named *Oreopithecus*, it is unfortunately unrelated to the human species. Its skeleton seemed to be indicative of the upright position, but this was later explained by the fact that this animal, like the gibbon, swung through the trees, its feet never touching the ground.

Looking back along the tracks, other minor stations can be seen, but they were probably not linked to Laetoli. Like the so-called *Ramapithecus*, they probably formed part of another network.

Actually, there are a number of "ramapithecine" stations, rather clearly connected to one another. They include the animals found in India, China, Pakistan, Greece, Turkey, Hungary, and Kenya. Despite

5. Reconstruction from the remains found of a *Proconsul*. It lived between 14 and 23 million years ago.

different names, they all belong to the same group, which lived from 7 to 17 million years ago. The most famous names are *Ramapithecus* and *Sivapithecus,* named after the two Indian divinities Ramayana and Sivah. (We won't burden the reader with the complicated "distinctions" within the group, which would place the species *Ramapithecus* under the genus *Sivapithecus*.)

For many years it was believed that the Ramapithecines belonged to the lineage leading to *Homo*, but various studies have now ruled out that possibility (they are ancestors of the orangutan).

Can any other stations be seen on the horizon? Of course. But the rails linking them are missing. The difficulty in reconstructing the distant past is due to the fact that the "genealogical trees" pieced together by experts include a number of animals (officially defined as "dryopithecines") that seem to diverge in quite different directions and have uncertain links between them.

What interests us is the upright position: was there an evolution toward the upright position among these many pre-hominids? This question is very difficult to answer because, as far as we know, these beings either walked on all fours, or climbed and swung from trees, or a combination of the two. So, it is not easy to identify our direct lineage (this discussion will be continued in the next chapter).

Are there any other clues?

The Y

Well, there is one, but it is very tenuous.

The dryopithecines seem to show an evolution toward a more modern set of teeth. In particular, the surface of their lower molars showed a Y-shaped pattern. We have this Y-shaped pattern on our molars and so did Lucy; it seems to be a kind of marker, a tracer leading back through time.

This Y does not allow us to identify which dryopithecines were our ancestors (because the characteristic is found throughout the group), but it does allow us to go further back in time. Where does it take us?

We know that one of the most ancient representatives of the dryopithecines, the *Proconsul* (which lived between 14 and 23 million years ago) also had molars with a Y. Being close to the top of the pyramid in this group, it may have been our ancestor (as well as an ancestor of the gorilla, the chimpanzee, and the orangutan).

How far back can the Y take us? All the way to the *Propliopithecus*,

6. Reconstruction from the remains found of an *Aegyptopithecus*. It was the size of a fox and had a brain the size of a ping-pong ball.

which lived 30 to 35 million years ago, but all we have to go on are its mandible and teeth. We know more about its close relative, the *Aegyptopithecus*.

The *Aegyptopithecus*, found at Fayum in Egypt, which lived between 31 and 34 million years ago, is for many reasons considered not only a close relative of the *Propliopithecus*, but also an ancestor of the *Proconsul*. So, it may have been our ancestor as well.

What was it like?

Well, it was quite different from us—it didn't even look like an ape. It was more or less the size of a fox, with a long snout, a long tail, and a brain that measured 27 cubic centimeters (the size of a ping-pong ball). It was a land animal, a quadruped, an agile climber, and a herbivore.

The very idea of having a quadruped as an ancestor is already somewhat embarrassing, but coming upon this kind of animal in our evolutionary chain is definitely disconcerting. The feeling can only increase the farther back we go; at 60 million years ago, we come upon forms of life that resemble squirrels.

Continuing (to 70 million years ago), we find an animal called *Purgatorius*, a kind of small rat. It lived in the trees and ate leaves, bark, and grains. It had a long snout and 44 teeth. Many scientist believe that this is the most distant station along the evolutionary railway line.

Although the reconstruction of the *Purgatorius* is based exclusively on remains of the teeth and the jaws, it is generally considered valid. Most paleontologists agree that *Purgatorius* (so named because it was found for the first time on Purgatory Hill in Montana) is the most distant ancestor of human beings and apes.

So, going back 70 million years in time from one generation to the next and imagining the sequence speeded up on film, we would see our ancestors change appearance. They would first become hairier, then more stooped, then quadrupedal, then smaller and more and more like a squirrel. Finally they would be like a tree shrew.

The *Purgatorius* was a contemporary of the dinosaurs. It was one of the first mammals (it descended from the reptiles) and witnessed, without realizing it, the extinction of these conquerors of the land. Little did it know that it was to give rise to a line of beings that would slowly lead to a new and completely different species of dominators: *Homo sapiens sapiens*.

The African Divide

Of course, this evolution which preceded the evolution of human beings took place in an environment that was in continual flux, and this profoundly affected its various stages. At this point, it might be a good idea to paint a rough picture of the background: the scene of the crime, so to speak.

According to the reconstruction of Yves Coppens, two important events took place in Africa (which was the site of all the crucial stages of human evolution).

The first took place 16 to 17 million years ago. Due to the movement of the earth's crust (so-called plate tectonics), Africa and Arabia moved northeast and closed off a large body of water connecting the Mediterranean with the Indian Ocean. This established a natural land bridge, similar to the one that exists today, making it possible to pass from Africa to Asia or Europe.

The *Proconsul,* or some similar form, immediately took advantage of the situation and started to travel, spreading its descendants first in Europe and later in Asia.

In Coppens' opinion, an African descendant, perhaps the *Kenyapithecus* repeated the same operation around 14 million years ago, and

7. Reconstruction from the few remains found of a *Purgatorius*. It lived 70 million years ago, at the time of the dinosaurs, and could well be the founder of the mammalian line that led to primates and human beings.

this gave origin to the many forms of Ramapithecines scattered between Hungary, China, and Pakistan.

The second important event took place about 10 million years ago. Africa split longitudinally. (The process had already begun previously as a result of tectonic movements.) This resulted in the line, still visible on a map today, of great lakes extending north-south from Ethiopia to Mozambique.

The earth was heaved up along this divide (the "Rift"), volcanic activity increased, and important climatic changes took place. While the land to the west remained covered with tropical forests, changes in air currents caused the area to the east to become increasingly arid.

Coppens feels that these climatic differences in the two areas caused different adaptations. In the forest, the forms that we are familiar with today gradually developed (apes, gorillas), while in the arid savannah area of eastern Africa, the first forms of hominids appeared.

The event we are interested in—the divergence between the ancestors of the apes and those that led to the hominids whose footprints were found at Laetoli—took place in that area in a manner parallel to the geoclimatic changes.

Unfortunately, as has been mentioned, that period coincides with a "black hole" in fossil remains. That means that the pieces needed to complete the puzzle are missing, at least for now.

What other line of investigation can we follow in order to find out more about this period? Is there any way we can study the period of that crucial bifurcation without fossil remains?

The Molecular Clock

Some people have taken a completely new tack. Just as DNA is now being used in judicial inquiries to identify guilty parties, so is biochemical analysis now being applied in paleontology to a few cells left at the "scene of the crime" (blood, hair, etc.) in an attempt to reveal the past.

The original idea was rather bold: to try to find traces of proteins in fossils millions of years old. A crazy idea and, in fact, that's what people thought of Professor Jerold M. Lowenstein of the University of California at San Francisco when he announced his intention to look for proteins in ancient fossils. Lowenstein hypothesized that the slow

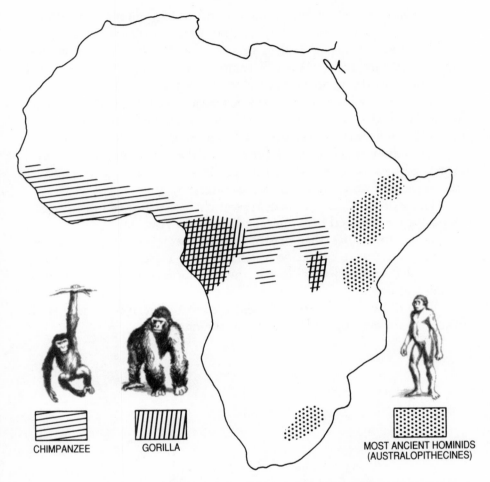

CHIMPANZEE   GORILLA

MOST ANCIENT HOMINIDS
(AUSTRALOPITHECINES)

8. Approximately 10 million years ago, a vertical split in eastern Africa (the Rift Valley) caused a climatic division. Gorillas and chimpanzees developed in the forests to the west (areas indicated by lines), while hominids developed on the savannah to the east (the dotted areas indicate where fossil remains have been found).

substitution of molecules transforming bone into rock could, in some cases, be incomplete. If so, perhaps traces of organic substances (such as collagen) would still be present in spite of the long time passed.

His hypothesis turned out to be right. After grating certain fossil fragments and pulverizing them, Lowenstein subjected them to radioimmunoassay, a well-known technique by which, among other things, the affinities among living species can be determined (for example, how "far" a human being is from a mouse, a horse, or a chimpanzee on the evolutionary scale). We will return to this subject in chapter 11, which deals with genetics and genealogical trees.

Suffice it to say here that this kind of research indicates that the divergence between the ancestors of the chimpanzee and those of the hominids took place between 5 and 7 million years ago.

So, radioimmunoassay confirms that these common ancestors lived in that period, as suggested by the geoclimatic transformations and the paleontological finds (although still incomplete).

Therefore, one of the most typical characteristics of our species—bipedalism—appeared between 5 and 7 million years ago.

Since the beings that existed before that time (Ramapithecines, *Proconsul*, etc.) were quadrupeds, and those that existed after that time (at Laetoli) were perfectly bipedal, the transition must have taken place during that time.

How did it occur? And why did creatures that lived mainly in the trees start to stand up and walk on the savannah?

3

The Birth of the Bipeds

Too Many Hypotheses

The emergence of bipedalism is, of course, one of the most important milestones in the evolution of the human race. It is also of great emotional significance, as it is often taken as the magic moment of transition from animal to human being. Actually, as will be seen later, things are much more vague (and complex).

This chapter will try to explain them in a simple manner, but some points call for detail which we hope won't overburden the reader. We will then continue our general discussion in the next chapter.

Many hypotheses have been formulated about the origin of bipedalism. Some are plausible, others less so. To mention just a few: our ancestors adopted the erect position in order to carry food, to use tools, to see over the tall grass, to gather berries hanging at the top of the shrubs, to seem taller and more fearful to adversaries, to show their sexual attributes, to be able to throw stones, to stand up in water, to expose their bodies less to the sun, and for mechanical and energetic reasons due to a change in diet.

The list is long, but none of these explanations can be confirmed at the moment. The reason behind bipedalism is not yet clear. Some hypotheses have, however, been ruled out in recent years, such as the one that postulates that hominids stood up in order to be able to use tools. The most ancient implements only appeared two million years after the emergence of the upright position. So, it is more likely that their use is a consequence than a cause. Before using stone tools, hominids probably used clubs, lianes, or unchipped rocks for a long period of time.

Alan Walker, of Johns Hopkins University in Baltimore and a close collaborator of Richard Leakey in Kenya, explains:

> Evolution cannot be seen as pressure to toward more and more perfect beings, but merely as a series of compromises. Bipedalism is clearly not very favorable from an energy point of view: quadrupedalism is much more so. Therefore, there must be other advantages. Many theories have been advanced in the past, but they all have one defect: none of them can be proven.

Alan Walker is right. Let's try to leave theories aside (for now), too, and start with the facts, as every good investigator should. Simple consideration of facts can provide a lot of information.

One fact that everyone agrees on is that bipedalism was not achieved overnight. It is impossible that an individual started to walk in the upright position as a result of a genetic mutation, for example. Why? Because it would be too complicated. As mentioned earlier, the transition involves changes in the structure of many bones, tissues, organs, muscles, joints, and nerve endings, which cannot all have taken place at the same time.

Exactly which changes were required for the transition from the quadrupedal to the bipedal position? Although clear, the next paragraph is rather technical. It can be skipped by the impatient reader.

A Very Complex Transformation

Let's start with the brain. All we have to do is look at a human being and at a dog to realize that profound changes have taken place in the transition from the "horizontal" to the "vertical" position.

The human skull has shifted its point of attachment in order to sit at the top of the spinal cord like a ball balanced on a stick. The foramen magnum (the opening in the skull through which the cranial cavity communicates with the vertebral canal) has shifted underneath the skull, whereas in the dog and other quadrupeds, in general, it is situated in the back. This characteristic is very useful in human paleontology in determining whether a creature walked in an erect position

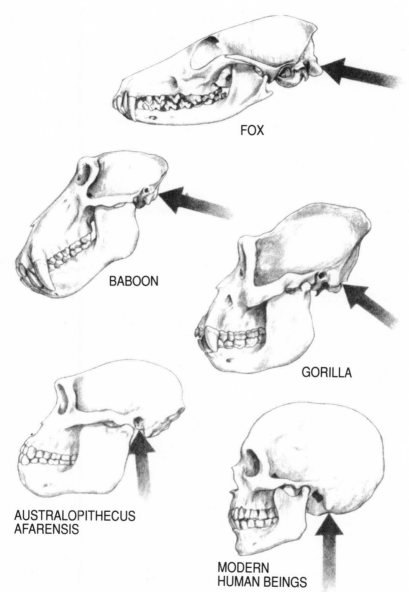

FOX

BABOON

GORILLA

AUSTRALOPITHECUS
AFARENSIS

MODERN
HUMAN BEINGS

9. In man the spine is perpendicular to the skull, with the skull resting on top of it like a ball on a stick. In other animals, the position of the skull with respect to the spine varies from nearly horizontal in the fox, for example, to inclined in the baboon and gorilla. In *Australopithecus afarensis* it was almost vertical. The upright position is, therefore, also the result of a shift in the occipital foramen.

or not. (The foramen magnum of the chimpanzees that walk on their knuckles is situated halfway between that of quadrupeds and that of human beings).

But the new position of the skull also has other consquences: for example, adjustment of the muscles at the back of the neck. And of the bones: since such strong muscles were no longer required to hold up the skull (which is in equilibrium), the bony crest on the back of the skull found in all quadrupeds and the bony spines of the cervical vertebrae which offer support for the muscles at the back of the neck were no longer needed.

Analogous transformations took place in the pelvis. In order to provide a broader attachment area for the gluteus muscle and a more effective position for the muscle on the femur, the "iliac blades" swiveled forward and broadened, while the rest of the pelvis shortened.

The entire structure took on the typical basin-like shape found in all hominids but absent in apes. This shape is extremely useful in supporting the internal organs, which began to sag in the vertical position.

Weight also posed a new problem: with only two points of contact, all weight was borne by the spinal cord and the legs and feet. So the spinal cord developed a series of curves (kyphosis and lordosis) to provide flexibility in walking erectly. The sacrum, which bears the weight of the entire trunk, became shorter and broader, while feet became arched (using the principle employed in the suspension of trucks) to cushion contact. It is well known that people with flat feet (that is, with an inadequate arch) tire more easily when they walk.

Feet also underwent other important modifications. Besides the natural enlargement of the heel and one bone in particular—the astragalus—situated directly under the tibia, which bears the entire weight of the body, another typical difference between human beings and apes emerged: the big toe became aligned with the others (unlike the thumb of the hand). Having lost its prehensile function with the move from the trees to solid ground, the big toe took on a fundamental function in bipedalism: thrust. In order to do so, it had to line up with the other toes and enlarge in order to bear the stress produced by walking or running (just think of the strain our toes are subjected to during a 100 meter race).

And finally our legs. Our femurs are longer and larger than those of a gorilla, because we are the only primates that use them so intensely.

In biological terms, we are animals that are specialized in walking up-right, just as birds are specialized in flying and kangaroos in jumping. Therefore, the femur had to undergo many changes, the most typical of which is its inclination with respect to the axis of the body. In order to be able to walk more efficiently, the knees drew closer together, forcing the femurs to stand at an angle.

There were also numerous other changes—in the ligaments, the discs between the vertebrae, the circulatory system (pumping and return of the blood), the nervous system, and so on.

Intermediate Changes and Compromises

As we said earlier, these changes cannot all have taken place at the same time. They are not merely adaptations, they are real genetic muta-tions which accumulated over time and gradually produced a functional and harmonious structure.

Brigitte Senut, of the Institute of Paleonotology of the Museum of Natural History of Paris, specializing in the problems of locomotion, puts it this way:

> These evolutions are seldom rapid; they take millions of years. There were probably a number of "attempts" during the course of evolu-tion, perhaps in different places. All together, these changes result in structures that are not very useful initially, but which become more so as certain transformations in climate and the environment take place.
>
> *Therefore it is difficult to understand exactly how the transition to bipedalism came about? Whether from the feet up or from the pelvis down?*
>
> We will probably never know. What we do know is that each animal adapts to its environment: if the environment changes, the animal reacts with the pre-adaptations it possesses; otherwise, it becomes extinct. We must not forget that each stage of the evolu-tion of the primates and of man caused an overall reaction be-tween the body and the brain: all locomotive orders or muscular orders originate in the brain and are sent to the body through the nervous system.

But are there fossils that attest to an intermediate stage, a kind of "compromise" in locomotion?

Yes. Even today, some locomotive compromises can still be found among the primates: the gibbon can walk on two feet, or on four, or it can jump and swing from branches, as can many other primates. These compromises are evident in the fossils. Too often in studying bipedalism, we forget to study not only human beings, chimpanzees, and Australopithecines, but also the earlier stages.

But there are practically no fossils from between 5 and 10 million years ago. . . .

Yes, but if one goes farther back in time, to 18 to 20 million years ago, for example, there are many. Outstanding are those of the *Proconsul*, which seems to have all the credentials for being the common ancestor of the great apes and human beings. In particular, the finds made by Leakey and Walker in 1984–85 contained complete skeletons and other material. It seems that these beings were already capable of rather extensive locomotive compromises: walking on all fours, hanging, climbing trees.

This opened up various evolutionary possibilities?

Exactly. This locomotive melting pot probably made the development of specialized systems possible. Another example: in a find dating back approximately 15 million years, at Fort Ternan, Kenya, bones were found which still have not been identified (there are a number of possibilities). But in any case, they belong to an animal that was both a climber and a quadruped, and that is an important clue.

A Great Number of Bipeds in Nature

As Brigitte Senut says, it won't be easy to give an exact reconstruction of the sequence that led to bipedalism. But given the time required and the succession of generations and environments (exerting various kinds of "selective pressure"), this mechanism is a reasonable explanation of how things went.

However, one fact tends to be overlooked: human beings are not

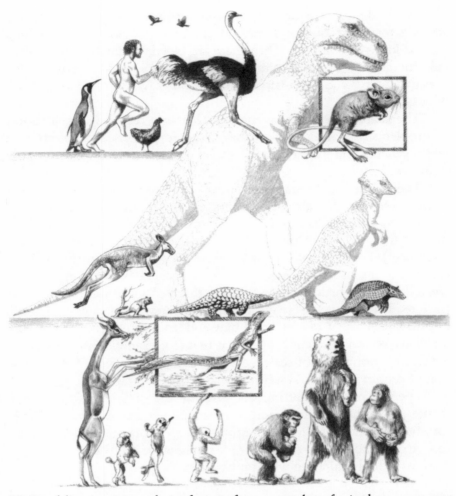

10. Bipedalism is not an exclusive feature of man: a number of animals are permanent bipeds (ostriches, penguins, birds in general). Others walk on two legs, using the tail for balance (kangaroos, scaly anteaters, armadillos, many types of dinosaurs). Still others are occasional bipeds (dogs, bears, monkeys, and basilisks, which can also run on water).

the only bipeds on the face of the earth. Nor were they the first. The same process took place many other times in nature, although in different ways, and gave origin to other species that walk on their two hind feet.

Many animals are capable of standing up on their hind feet and walking for short distances: cats, dogs, bears, horses, elephants, and various kinds of primates. Then there are bipedal animals that use their tails for balance (such as tyrannosaurs, basilisks, and kangaroos) and non-flying birds such as ostriches and penguins. When on land, all birds walk on two legs.

The adaptations of these animals obviously responded to a variety of problems; but in each case, they required processes of selection in which genetics and the environment worked together, as in the evolution of human beings.

Then again, some transformations and evolutions were far more complicated than the emergence of bipedalism. Just think of the evolution that led to flight: it is infinitely easier to stand on two feet and walk than to fly! And yet, even flight has been achieved in nature (and not just once!). Fossil remains start to show pre-adaptation to flight: here, too, the accumulation of mutations (that is, genetic "errors" in heredity) created so-called neutral characteristics, that is, "anomalies" which were tolerated because they were not harmful and were carried along through evolution as a slight burden until they suddenly proved useful. Or they may have served different purposes at first (for example, feathers, deriving from "anomalies" of scales, may have initially improved the insulation of certain reptiles and were only later used for gliding).

The fact remains that reptiles were not the only animals that took to the air: some fish and even some mammals finally achieved flight for short periods of time (sea robins) or permanently (bats). In any case, it definitely took more complex adaptations for a mouse to fly than for a pre-hominid to walk on two legs.

Pros and Cons

It is still not clear why pre-hominids, who lived in a difficult and competitive environment, evolved toward permanent bipedalism. Was the form of locomotion not dangerous and full of drawbacks? Walking on two limbs alone meant risking almost total immobility in case of injury or fracture, thus increasing vulnerability to predators and making it more difficult to procure food and water. It also meant moving considerably more slowly than other mammals: human beings are among the slowest animals of the savannah, be they prey or predator. No hundred-meter record holder can compete with a hyena, a lion, a jackal, or even a dog or a cat, let alone a gazelle or an antelope. And the savannah is not flat like the track in a stadium.

But bipedalism of human beings does seem to have some advantages. Some scholars believe—although it is still a controversial idea—that the erert position is energy-saving. But there may be other, more interesting advantages, as Professor Bernard Vandermeersch, of the Anthropological Laboratory of the University of Bordeaux, claims:

> It's true that human beings do not run fast, and that any quadruped can move much more quickly than they do. But human beings are great long-distance runners: no other mammal is capable of running for such a long time. Some populations still exist today in Australia and in Mexico that pursue their prey (some kinds of deer and kangaroos) until the animals collapse. This means that they run for hours on end. Human beings can stand that, animals cannot.
>
> Moreover, human beings can change speed much more easily than animals, which are adapted to minor variations in speed.

The upright position must have brought far more substantial advantages to justify its success among hominids. Advantages that more than compensated for the physiological drawbacks which today, after millions of years, still cause modern human beings trouble: slipped and flattened discs, sciatica, prolapse, varicose veins, and knee problems.

What were those advantages?

Although there is no evidence to support any particular hypothesis, one is of special interest. Interesting, stimulating, and informative, it involves the extremely important matter of reproductive strategy. Reproduction is an essential element of all life; the continuation of the species depends on it.

But what have reproductive strategies got to do with the upright position? We will soon find out by following a very subtle but fascinating line of reasoning put forth by Owen Lovejoy at Kent State University, an expert on locomotion.

The Hidden Estrus

Before going into Lovejoy's theory, some background information has to be given. One obvious fact, but one which we rarely notice, is that woman is the only mammal that does not go into heat. During the three days of fertility (halfway through the menstrual cycle), she provides no sign or odor to indicate that she is in that critical period.

Anyone who has a cat or a dog knows that during estrus (that is, "heat"), the females draw suitors from all over the neighborhood. In fact, the genitals emit an odor that is carried in the air and excites the males in the vicinity.

In mammals that live in groups, these sexual messages produced by the female provoke fights among the males for possession: the dominant male, who has already demonstrated his supremacy within the group in other territorial struggles, drives competitors away and fertilizes the female. This is the harem model, very widespread among mammals. It is the rule, for example, among gorillas, baboons, and sacred baboons (chimpanzees, on the contrary, are very promiscuous).

The dominant male is not always the only one to mate with the females. Among the baboons, for instance, some young males also copulate, but only during non-fertile periods, that is, when the females are not "in heat." When estrus begins, the leader becomes aggressive and keeps all rivals away because that is the only time when fertilization is possible. Estrus is, in fact, a manifestation of ovulation signaling the magic moment of reproduction and the transmission of the genes.

But once again, what has all this got to do with the evolution of bipedalism? There may be a connection.

First of all, why don't women have estrus? In other words, why don't they let males know that they are ready for reproduction? (It seems that only 30 percent of all women emit a very faint odor during the fertile period, which probably goes unnoticed by most men.) How far does this characteristic go? How did it develop?

Actually, it should be an evolutionary disadvantage: in nature, females that are in heat draw more males and as a consequence reproduce more (and we know that in nature neither the strongest nor the most beautiful nor the most intelligent succeeds; the creature that produces the most numerous and the most fertile offspring succeeds). Therefore, the lack of estrus would apparently be a negative trait.

But let's go back in time to a group of arboreal hominids living in the tropical forest, that is, to creatures similar to chimpanzees (living partly in the trees and partly on the ground) except for one major difference: they already assume a more upright position and run on the ground on two legs. (The primeval forest in Africa is not the impenetrable jungle often imagined; it is made up of large trees whose upper branches touch, but the large open spaces at their bases provide a suitable habitat even for some kinds of elephants.)

Now let's imagine that a female with a flaw is born into this group of pre-hominids: she has no estrus. What would happen? The dominant male would take no notice her and would not mate with her.

But this would not keep her from mating with some young male, without interference from the dominant male. A peripheral, unopposed kind of relation could grow up between this "lesser" male and the "forgotten" female.

Suppose that this female actually had an unsignaled ovulation (like women today), or rather a series of unsignaled ovulations throughout the year (not only in the mating season). This apparently "sterile" female could give rise to a line of descendants in which, statistically, the abnormality would recur.

Bipedalism and the Couple

This brings us to Lovejoy's theory. He hypothesizes that the absence of estrus in certain groups of hominids gradually led to the development of personal relations within the group that were no longer characterized by the dominant male, but based on pairing. They could even have been monogamous (as are gibbons, for example).

This would have led from simple episodic mating (mainly the prerogative of the dominant male in the rare periods of estrus) to a myriad of relations between males and females within the group. Relations could have become more frequent and more stable, because only those constantly close to a female could fertilize her, as ovulation was unsignaled and, therefore, unpredictable (as often happens in the human race).

Was this the prelude to "love"? That is, to couples established (and maintained) by means of certain permanent signals no longer indicating only estrus? Was this the beginning of other kinds of attraction, for example, to certain physical attributes like eyes, hair, and skin?

That is what happens in human beings. Our model, so different from that of other mammals, must have originated somewhere, for some reason. Can it be rooted in the ancient evolution of pairing, which already existed in pre-hominids?

Pairing is not that rare in nature: just look at birds. Most birds mate for life: both the male and the female take part in parental care and hunt for food (for themselves and for their young). In fact, if the male deserts the nest, half the offspring die; thus, permanent collaboration is a successful behavior.

Collaboration in pairing could have led to the sharing of food among pre-hominids, too. And this is the essential point which, in Lovejoy's opinion, led to bipedalism.

He hypothesizes that this personalized relationship led to a division of labor: the females cared for the young, while the males went out in search of food. This collaboration might have been particularly important when, as a result of climatic changes, the great forests of eastern Africa started to disappear and give way to the savannah.

The advantage of having their hands free to gather and carry food could have been an important reinforcement of the collaborative pairing model. And this model might have made it possible to have more children. Why?

Two Different Reproductive Strategies

In nature, there are two reproductive strategies: "r" and "k." The "r" strategy produces a lot of children without much concern for them: some of them are bound to survive (turtles, for example, lay an enormous number of eggs on the beach and then abandon them in the knowledge that some will inevitably be able to reach the sea, avoiding numerous perils and predators).

The other strategy, "k," is exactly the opposite: few children are born but are well taken care of. The large anthropomorphous apes (gorillas, chimpanzees, and orangutans) behave in this way, giving birth to only one progeny every five or six years. This low fertility rate is related to the fact that the mothers cannot count on the help of their partners: they have to look for food for themselves and for their young. This dual role—mother and provider—prevents them from giving birth to another offspring before the first is completely weaned (the absence of estrus during this period makes them temporarily infertile).

A collaborative couple, on the other hand, can generate more children (without having to wait five to six years) and can take equally good care of them. In evolution, having more children is definitely an advantage (if one can care for them adequately).

In the pairing model, the female would have been able to stay close to her young and the "home" (probably located in the trees), thus becoming the first "housewife" of human prehistory. Thanks to his upright position, the male would have had his hands free to gather and bring food "home."

Collective life could also have been affected by this model: formerly dominated by a male, the community gradually started to be made up of couples (in this way, the offspring had two parents rather than one). The decrease in competition could even have led to greater collaboration among the males, creating more cohesion in the group, against predators, for example.

One might almost say that it was no longer a group but a tribe. This could well have been a fundamental transformation for all of future development based on collaboration.

The greater care and protection offered by this model made it possible for the young to gain their independence more slowly, taking advantage of parental teaching for a much longer period of time. Looking at our species, this is the characteristic that has made cultural development possible: our young are extremely vulnerable and dependent for many years (they are actually unable to walk for a year or two), but that is precisely the period in which the fundamental maturation of the brain (which then becomes cultural) takes place.

It has been found that children who are abandoned at a very young age (or left on their own in the home) do not learn how to walk; they can only crawl. We *teach* our children how to walk. This is a demonstration of the importance of socialization in the history of our species.

In conclusion, Lovejoy basically sees the emergence of bipedalism as the result of a better reproductive strategy; only later did this model lead to the development of the brain, the use of tools, the demographic explosion, and the birth of culture.

From the African Lakes to *Swan Lake*

This theory of the emergence of bipedalism is, as mentioned earlier, pure hypothesis. We took the time to describe it here because we found it particularly stimulating and in some ways extremely plausible. One of the reasons for this is that it gives primary importance to the factor of "reproductive strategy," which is essential in understanding the history of evolution (and so often overlooked).

But investigators cannot go on theories. They need facts; therefore, as far as we know, things may have been quite different. The role of estrus may only be a "clue," a "guess," that could turn out to be misleading. Or it might have been responsible for other behavioral developments, but only at a later time, when the upright position was already a firmly established characteristic.

While agreeing with the fundamental reasoning behind this hypothesis, Tim White, paleontologist at the University of California at Berkeley, underlines that there is no proof to support it and adds, with the healthy skepticism of a field researcher, "Unfortunately I have never seen a fossil of estrus."

What we do know is that walking erect has been a success. Proof is that we walk in our parks, play soccer in the World Cup, and dance Tchaikovsky's *Swan Lake* on the tips of our toes. This evolution may have started in very ancient times among beings that were versatile in their modes of locomotion. We have seen how complex the transition to a permanently erect position was, but we have also seen how the upright position was and is used by other animals (and is therefore neither unique nor unexplainable).

What the footprints at Laetoli teach us is that this evolution was complete 3.7 million years ago.

Let's take a close look at those footprints one last time. We have seen where they come from; let's try to see where they lead.

We'll start by describing the hominids that left them: the Australopithecines.

4

The Unsettling Faces of the Past:
The Australopithecines

Traveling on the Savannah

Here we are at Laetoli once again. Analyses and laboratory tests have shown that these two or three hominids walked as we do, were slightly shorter than the Pygmies of today (between three and one-half and four feet), and were around nineteen miles from a huge volcanic eruption which had deposited a layer of ash on the entire area, transforming it into a lunar landscape.

What did they look like? Although insufficient remains—mandibles, teeth, and maxillary bones—have been found in the Laetoli area to be able to piece together a portrait, they are similar to those found in other areas, where a rather complete reconstruction has been possible. These other hominids have been discovered in some areas of eastern Africa, in particular to the north of Laetoli, in Hadar, Ethiopia.

It must be underlined that the finds are distant both in space and time. The remains in Laetoli date back 3.7 million years, while the fossils at Hadar range from 2.9 to 3.7 million years ago (dating varies dependng on the site and the dating technique used). The two localities are around 940 miles apart—the distance, as the crow flies, from Paris to Rome. But traveling that distance over one hundred thousand years would mean about 45 feet per year (less than two inches per day).

Given the similarity of the fossils, it is reasonable to assume that they belong to one species which spread throughout that area of eastern

Lucy in the sky with dia-monds.

11. A famous prehistoric twenty-year-old: Lucy, who owes her name to a song by the Beatles. Forty percent of the skeleton of this Australopithecine was found, providing paleoanthropologists with enough pieces to make a reliable reconstruction of these ancient hominids.

Africa, perhaps moving only very little in the course of a century. The species was well adapted to different climates: the arid savannah of Laetoli as well as the semi-forest of Hadar.

With due caution, it may be assumed that the Hadar fossils provide a good approximation of the hominids that left the footprints in Laetoli. The fossils found in Hadar are indeed extraordinary: one in particular—Lucy—is now famous throughout the world.

Lucy was found by Donald Johanson and Tom Gray in November

1974, in the Afar region of Hadar (94 miles northeast of Addis Abeba). During a routine exploration of an area already surveyed many times, Johanson suddenly noticed something on a slope. Going over to take a better look with Gray, he realized that it was the skeleton of a hominid, the most complete ever found from such an ancient time.

Certain characteristics suggested that it must be a female, so they named it "Lucy," after the Beatles song incessantly being played in their camp, "Lucy in the Sky with Diamonds." From that moment on, in her travels around the world, Lucy has been the subject of much talk and study. What are the results of these studies?

A Prehistoric Twenty-Year-Old: Lucy

Forty percent complete (and even more, considering that many missing bones can be reconstructed as mirror images of those that exist), Lucy has a typical female pelvis. She is presumed to be twenty years old: analysis of her teeth shows that her wisdom teeth had already erupted but were not yet worn down.

Height: 36 to 44 inches (slightly shorter than the smaller of the two individuals at Laetoli). These figures are the result of studies of the femur in relation to other bones. She was as tall as a six- or seven-year-old girl today.

Weight: around 50 pounds.

Cause of death: A sudden event. There are no signs of predation on her body, which means that she was quickly covered by sediment. It looks as if she fell into the water while standing on the banks of a lake or a river. Perhaps she had a stroke, or drowned, or was caught by a flash flood.

Brain: very small. Only a few fragments of the skull were found, but a reconstruction based on other findings indicates that the skull was no bigger than a large orange.

Posture: definitely erect. A perfect biped (but probably also a good tree climber).

Body structure: already similar to ours. The arms are still a little long, but not as long as those of a chimpanzee. The relation between legs and arms seems to be somewhere between that in human beings and that in chimps: the humerus is 85 percent the length of the femur (in human beings that ratio is 75 percent; in chimpanzees, the humerus is longer than the femur).

Face: extremely primitive. The jaw and the teeth still have an archaic structure. Her face presents surprisingly ape-like features.

Donald Johanson comments: "Tim White and I were shocked when we started to analyze the pieces. We immediately realized that the jaws were extremely primitive, not human with ape-like tendencies, but ape-like with human tendencies."

Lucy finally offered paleontologists a number of important pieces of the mosaic: up to that time, only occasional fragments had been found and reconstruction, although reliable, was based on supposition. But Lucy was like a "do-it-yourself kit." They could finally check the validity of their assumptions. It would be as if, a million years from now, while digging in the strata from our era, archeologists were to come upon a piece of carburetor, half a steering wheel, a fragment of tire, etc., and were to try, on the basis of that information, to figure out what an automobile looked like. Discovery of an almost complete car would confirm their assumptions about its structure and some of its performance characteristics, and would allow the archeologists to place new pieces more accurately.

Lucy, discovered in Afar, was named *Australopithecus afarensis.*

Was Lucy the only form of hominid that lived at that time? The genus *Australopithecus* actually includes numerous other hominids that lived after Lucy and before the emergence of the genus *Homo.*

Australopithecus: Variations on a Theme

The name *Australopithecus,* coined at the time of the first findings in southern Africa (the first was in Taung, South Africa, in 1924), literally means "southern ape." For investigators today, the Australopithecines are a strange family with unclear family ties. The period of their existence extended over a very long time—at least two and a half million years (from approximately 1.2 to 3.7 million years ago) in various areas of south and east Africa.

This family will be discussed in more detail in this chapter, since it seems to represent the transition stage toward increasingly human forms of life. These still rather mysterious creatures, halfway between man and ape, have sparked the imagination of writers, artists, and directors (remember the first scene in *2001: A Space Odyssey?*). Let's see what accurate scientific study has revealed.

Four basically different kinds of Australopithecines have been identified: some are frailer, some stockier, some taller, and some shorter, with different dentition at times. Each of these Australopithecines has been given the name of a species: the oldest is known as *Australopithecus afarensis* (Lucy); the South African form, *Australopithecus africanus*; the two stockier forms, *Australopithecus boisei* and *Australopithecus robustus*.

The figure on the following page indicates the places in which the various forms were found and the periods in which they presumably lived.

As can be seen from the map, almost all finds lie along the so-called Rift Valley, that is, the African divide mentioned previously.

They lived in a period that spans approximately 2.5 million years (from 1.2 to 3.7 million years ago). That is an extremely long time—it is practically the same length of time that separates the Australopithecines from us. We have only to look at the changes that have taken place in the last 2.5 million years (not only in our evolutionary line, but also in that of many other mammals) to realize that the genealogical branchings from the Australopithecines must be complex indeed.

We will return to this subject later: there are many "trees," but there are also many divergences among the experts. For now, let's consider the fact that, despite differences, the remains of the hominids living in that long period show common characteristics. Let's list them briefly.

Common features of the Australopithecines are:

1) *Body:* adapted to bipedalism (even if some, like the *afarensis*, were probably good climbers as well).

2) *Brain:* very small. Calculated in relation to the size of the skull, the volume of the brain varied from 400 to 550 cubic centimeters (equal to half a liter). Thus, brain size of the Australopithecines increased relatively little throughout their two million-year evolutionary period (in *Homo habilis*, their successor, the volume of the brain increased rapidly to 600–800 cubic centimeters).

3) *Face:* still very ape-like, with pronounced cheekbones and a projecting jaw (seen in profile, the lower part of the face—the jaw and the teeth—juts forward, like that of apes).

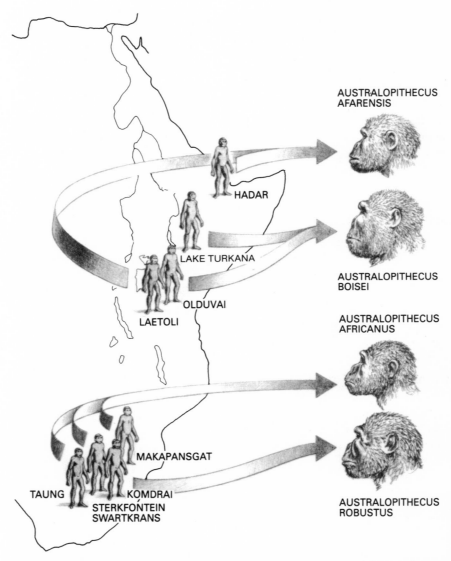

12. The distribution of Australopithecines in Africa, on the basis of the fossil remains found.

4) *Teeth*: evolving toward ours, but still very far removed. The enamel of the molars was incredibly thick (how fortunate for them), but also subjected to much wear.

5) *Diet:* study of the teeth shows that their diet was basically vegetarian, although varied, as will be seen later.

6) *Sexual differences:* pronounced dimorphism, that is, males were considerably larger than females. This is a characteristic of many primates, particularly chimpanzees and gorillas. (In our species, males are also generally slightly larger and more robust. In *Australopithecus boisei*, dimorphism is so pronounced that the first finds were thought to be of different species, rather than different sexes.) Dimorphism is a delicate question, however, since there may be considerable differences between individuals and groups; only a larger number of findings will make it possible to assess these differences statistically.

These common characteristics are, however, offset by a number of differences.

It should not be forgotten, in fact, that these hominids lived in very different environments and climates. Biology teaches us that living creatures are molded by natural selection in the environment in which they live. In particular, the availability of certain kinds of food creates specializations which are reflected in body structures, as we will see shortly.

In the meantime, let's take a look at the succession of Australopithecines (their "chronological tree," so to speak) to see how we can somehow order the material found to date.

Australopithecus Afarensis: Overture

It is commonly acknowledged that *afarensis*, who lived from 2.8 to 3.7 million years ago, is the most ancient Australopithecine. Some even more ancient fragments, dating back four million years or more, are also attributed to the same line—Lucy's.

What does scientific analysis reveal? Examination of the teeth shows that *afarensis* used their incisors a lot. This adaptation is typical of fruit-eaters. Why?

It is now known that teeth are an excellent indicator of the way in which food is chewed and eaten. For example, lions have long canines with which to bite their prey and strong molars (carnassials) to cut and tear bones and meat, but they have small incisors.

Rodents (such as mice, rabbits, and beavers) have strong incisors that grow continuously to make up for the wear caused by gnawing. If their jaws were unable to move, their teeth would continue to grow, turning into long curving fangs, like those of the wart hog. Cattle have large molars (which also grow continuously) to grind grass; the molars of elephants resemble millstones.

In the absence of large molars or canines, the well-developed incisors of *afarensis* would suggest an adaptation to the peeling and biting of fruit. (There is some evidence that *afarensis* was occasionally omnivorous.)

But a creature that eats fruit must obviously be able to climb trees. Did *afarensis* have that ability? Studies of the hands and feet of *afarensis* indicate that they were not straight, like ours; they were curved, and this is a typical feature of arboreal animals. But since we now know (from the footprints at Laetoli) that *afarensis* were fully bipedal and thus adapted to living on the ground, we can conclude that they alternated walking with climbing trees, where they found some of their food and perhaps shelter for the night (although there may have been types which were more specialized in one or the other).

This is just one example of how careful and combined analysis of the teeth, the hands, and the feet can provide information on the behavior of these ancient hominids: land dwellers, fond of fruit but probably omnivorous, still adapted to living in the trees, and capable of living in various semi-forest and tree-covered savannah environments. This ability to adapt to different climates may have allowed them to live without undergoing virtually any transformations for a million years.

Little else can be said about *afarensis* without slipping into uncorroborated hypotheses; it is very difficult to reconstruct events that occurred not thousands, but millions of years ago.

Afarensis most probably lived in groups (as do gorillas and chimpanzees). Discovery of at least thirteen individuals who died together in Hadar, site 333, seems to point in this direction. But there is much controversy and discussion about the thirteen individuals: some experts feel that they include two forms of hominids—a more primitive one and a more modern one—that lived at the same time.

More finds are required to support and confirm the hypotheses that have been made. Unfortunately, the current political climate in the Hadar area makes further investigation very difficult.

South Africa: A Hundred Caves for a "De Profundis"

Australopithecus afarensis disappeared around 2.8 million years ago. At the same time (perhaps a little earlier), new creatures, Australopithecines with similar features and a few differences, appeared on the scene.

Were they relatives? Descendants? There is no way we can know. All we do know is that these new Australopithecines, called *africanus*, have been found only in South Africa, over 1,800 miles away. Strangely enough, almost all remains have been found in caves although the *africanus* was not a cave dweller. Paleontologists have recently been able to account for this.

The area is the Transvaal region, rich in diamond mines. The place where these hominids were found is calcareous terrain, full of caves: a karstic area, in which vanishing streams generated by rain water have formed underground networks of tunnels and caves. This karstic area is full of pits that are partially filled with sediment and fossil remains—silt, leaves, and skeletons—carried there by the streams.

But what were hominids and animals doing in that area? The hypothesis is this: Although there were not many trees in the area at the time (certain evidence leads to this conclusion), the humidity close to the pits might have favored their growth there. Now, leopards like to climb trees to "park" and eat their prey at leisure out of reach of hyenas and vultures. So animal and hominid remains may have fallen down from the trees and rolled (or been washed) into the caves below.

This explanation is backed by examination of one of the "victims": the skull cap of a young Australopithecine (*robustus*) bears unmistakable traces of the fangs of a leopard, left while hauling its prey into a tree. After 2.5 million years, even minute signs like these can lead to the reconstruction of a dramatic event which took place on the South African savannah.

Other explanations are equally plausible: hyenas may have dragged their prey into the caves that have openings above the ground; or some Australopithecines may even have lived in the caves occasionally and been overwhelmed there by predators (this still happens at times to baboons in the area).

But what did this *africanus* species of the Australopithecines look like?

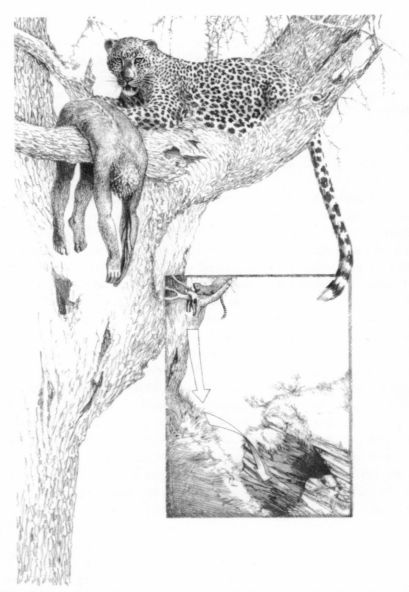

13. One of the predators of early hominids: the leopard. Unmistakable traces of leopard's fangs were found on the skull of an *Australopithecus robustus*. Seized by the skull and dragged to the top of a tree, the skull of the *robustus* probably rolled into a cave later, where it was found.

Australopithecus Africanus: Adagio, Senza Strumenti

Physically, they did not differ greatly from their *afarensis* "relatives." They were a little over four feet tall and weighed approximately sixty pounds. They still had very ape-like features with marked prognathism (that is, a very protruding jaw) and prominent cheekbones. *Africanus* also had a very pronounced brow.

In other words, if an *africanus* were found walking the streets of a city today, it would immediately be captured and taken to the zoo, despite the wonder over its bipedal capacities.

Hundreds of fragments of the remains of different individuals have been found—mostly teeth. Examination of the teeth has shown that the diet of *Australopithecus africanus* was probably somewhat different from that of *afarensis*: the incisors are smaller and the molars larger. They probably ate plants (a little harder to chew than fruit) and occasionally some smaller prey; perhaps even some meat torn from carcasses left by other predators.

The age at death can also be calculated from the teeth. After studying more than four hundred teeth of various kinds under the microscope, one researcher was able to establish the kind of dentition and the degree of wear. The average age turned out to be twenty-two years: death came early on the African savannah. These data correspond with the results of studies made in other areas on other individuals.

Was *africanus* more intelligent than *afarensis*? It is impossible to tell from the size of the brain (the volume of the brain was slightly larger, though, around 430–500 cc). Some specimens, such as the famous Taung skull (the first hominid found in Africa in 1924, unexpectedly brought to light by blasting in a rock quarry), show a potential for greater volume. The skull found in Taung has a cranial volume of 500 cc and is of a five- or six-year-old child; this would correspond to a volume of 600 cc in an adult. It must not be forgotten, however, that this skull is from the last period of *africanus*, about two million years ago.

From studies of the fossils he gathered in the vicinity in the 1920s and 1930s, Raymond Dart hypothesized that these hominids used bone tools. Not only that, he believed them to be great baboon killers: along with the remains of hominids, some of the caves contained large numbers of baboon skulls, crushed on the left side, as if violently struck by a blunt object wielded by a right-handed being. Dart was convinced that after killing the baboons, the hominids drank their warm blood.

More accurate study has disproved all these hypotheses. Like good investigators, some researchers systematically studied the dens of hyenas and found that the remains of their food show the same teeth marks and fractures as the bones found in South Africa; therefore, they were not tools but simply fragments of bone that had been eaten by hyenas.

Nor did *Australopithecus africanus* kill baboons. The animals were probably dragged into the caves by the hyenas. As for the battered skulls, the most likely to be explanation is that they were crushed by sediment layers.

Thus, *africanus* was probably not the bloodthirsty individual Raymond Dart made it out to be; on the contrary, it may well have been very peaceful and gentle. Why? Because its teeth, like those of *afarensis*, indicate that it was a vegetarian, more likely to be prey (for leopards, as was *robustus*) than predator.

It is probable, therefore, that *afarensis* walked the savannah not with the bearing of a killer but with a timid demeanor, worried about making unpleasant encounters and eager only to find its favorite plants, perhaps some eggs and small rodents, and, of course, water.

Australopithecus Boisei: The Nutcracker Suite

Two extraordinary actors, who walked the African savannah between 1.2 and 2 million years ago, must now be introduced: *Australopithecus boisei* and *Australopithecus robustus*.

As reconstructed by experts, they look like something halfway between an intelligent gorilla and a groggy boxer. What were they like? Although they lived far apart (the *boisei* in eastern Africa and the *robustus* in southern Africa), they look like close relatives. Imagine going back to the savannah to observe them at a distance, without being seen.

This is what an eyewitness might report.

About fifteen *boisei* are crossing a hilly area of yellow grass dotted by the occasional shrub. Upon reaching a valley between two hills, the group scatters. Some of the hominids stop to pick berries, others bend down to dig tubers from the ground. Gazelles and giraffes can be seen grazing in the distance. There may be some lions around.

The climate is very dry. There are few trees, some acacias and shrubs.

14. *Australopithecus africanus* were probably very peaceful creatures like the other Australopithecines. They were omnivores rather than predators, intent mainly on seeking out their favorite plants, some eggs, and perhaps a small rodent now and them.

A large blue lake lies not too distant; the *boisei* are probably headed for the water.

They are not very tall: just over four feet. The females, with their children beside them, are smaller. Their bodies seem to be dark and covered with hair, but we can't really tell from our vantage point.

They all look particularly stocky. They have a large head, prominent cheekbones, a flat nose and face. Many of them are chewing the tubers and fruits they have gathered, which are very tough and have to be chewed at length. These hominids have adapted well to a difficult environment and manage to survive on food that we would not even be able to digest.

With each movement of the jaws, the enormous temporal muscles bulge. These muscles, used in chewing, are so powerful that they reach the top of the skull where they unite and attach to a kind of bony crest. In this way, they resemble gorillas.

The architecture of their skull and their teeth was dictated by the need to chew their food for a long time. In fact, the teeth—not the incisors and the canines (which are small in proportion), but the molars, which are real grindstones—are as strong as nutcrackers. Just think that the crowns of our molars are normally one centimeter long; those of the *boisei* were sometimes as long as two and a half centimeters. That means about five or six times the volume of ours. The heads of these hominids have been compared to mills.

In this architecture, the brain was set well back: there was almost no forehead and the brow was so accentuated that it practically formed a visor over the eyes. Yet, despite this ape-like appearance, the brain of *boisei* measured 550 cc: much larger than that of the anthropomorphous apes. Not only that, the sulci and convolutions of the brain (calculated indirectly from the impressions on the skull) suggest a much more complex design.

Two small eyes glitter from under the arched brows. How intelligent were they? It's difficult to say. They still looked very much like apes. Their vegetarian diet leads to the assumption that they had not yet developed the ability for strategic thinking common to most predators.

Suddenly they all freeze and raise their heads. Danger seems to be threatening. The mothers clutch their children as the group draws together. Unintelligible calls can be heard. Some raise broken branches to be used as clubs. But it's a false alarm and the *boisei* soon continue on their way over the hill and disappear from sight.

15. *Australopithecus boisei* (which takes it name from C. Boise, the English patron of American origin, who helped Louis Leakey in this research) had immense molars and was literally a walking mill.

The "Black Skull": *Largo, Fortissimo*

Who the descendants of the *boisei* were has not yet been clearly established (there are a number of hypotheses), but a rather probable ancestor was recently found near Lake Turkana, Kenya. An amazing black skull (classified as KNMWT-17,000) dating more than half a million years further back (that is, to 2.5 million years ago) was found by Alan Walker. Some people, like Johanson, have proposed to call it *Australopithecus aethiopicus* because of a fragment of a mandible found previously in Ethiopia which seems to belong to the same type of hominid. A walking millstone, this individual had a large crest and an aerodynamic maxillary structure, designed for grinding tough food.

The Black Skull is, in a certain sense, a sign of its times. The appearance of these hominids coincides with a major climatic change which had important repercussions on the evolution of the fauna and the flora in

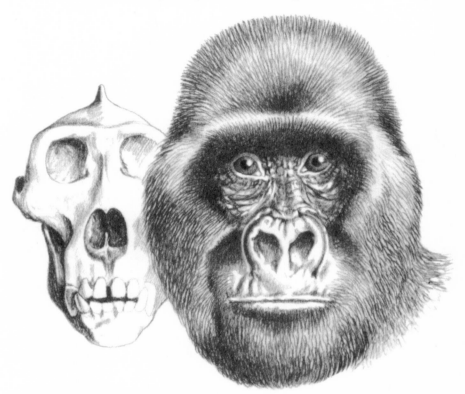

16. The bone and muscle structures of the head of a gorilla are very similar to those of *Australopithecus boisei*. In particular, both have a bony crest at the top of the skull to which the masticatory muscles attach.

that part of Africa. For reasons that are still unclear, but somehow related to the expansion of the polar icefields, the climate became much more arid and many tree-covered areas disappeared. Therefore, individuals "equipped" for the kind of food available in dry areas, such as tough plants with less nutritious leaves that needed to be eaten in large quantities and chewed for a long time, became dominant.

This is the kind of environment in which the Black Skull, the *boisei*, and the southern African *robustus* lived: all were walking mills, able to get the most in nutrition and calories out of the fruit and plants available, thanks to their ability to grind them with their teeth.

Let's move some 1,900 miles south now and visit the last member of this collection of Australopithecines: *robustus*, who lived in the savannah of South Africa near Swartkrans. This is one of the places in which the most numerous remains of *robustus* have been found—125 individuals. Thanks to these finds, some of the characteristics of this hominid can be reconstructed (the reconstructions of the *boisei* are based on data obtained from the remains of seventy individuals).

Let's return to our eyewitness.

Australopithecus robustus: Andante, ma non troppo

The landscape here is not much different from the one where the *boisei* lived: an arid savannah with little water. The animals are the same; but the area is still roamed by the legendary saber-toothed tiger, a predator similar to the lion, endowed with two ten-inch fangs with which it seizes and guts its prey.

In the distance, we see a group of *Australopithecus robustus*: about fifteen males, females, and children. They take small steps, rocking slightly in an ape-like way. In fact, their big toe is smaller than ours, giving them less leverage. Body weight is also distributed in a different way.

They strongly resemble their cousins, the *boisei*: they are the same height and weight and have the same brain volume. *Robustus* also have a large, flat face with enormous cheekbones and powerful muscles for chewing, but their features are less accentuated. Seen from afar, *robustus* are difficult to distinguish from their east African relatives.

Do they belong to one family which separated through successive migrations? Some people think so. Others believe that this is one of those strange "evolutionary convergences" which are frequently found in nature. In fact, two species (at times quite different from one another) often take on similar characteristics as a result of adaptation to similar environments. Many examples can be given: the jaguar of South America, which is identical to the African leopard, although it evolved on a different continent; or the Australian marsupial wolf (recently extinct), which resembled the European wolf despite its descent from a completely different evolutionary line of mammals, that of the marsupials.

Something similar may have occurred with *Australopithecus boisei* and *Australopithecus robustus*; that is, the environment, the kind of food available, and the climate may have shaped two similar models. Or are they variations on one form that extended vertically throughout eastern and southern

17. *Australopithecus robustus*, the remains of which have been found in southern Africa, was surprisingly similar to *boisei*, which lived at the same time thousands of miles to the north. A question of migration? Or just an evolutionary convergence as often happens in nature?

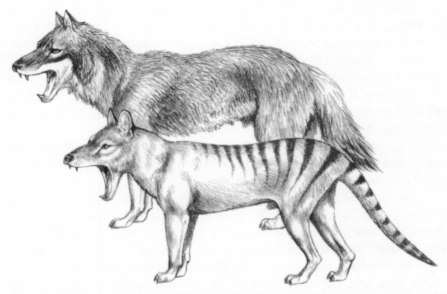

18. A European wolf and a Tasmanian wolf. They look like relatives—variations of the same species—but actually these two mammals are genetically distant: one is a placental, the other a marsupial. Environmental pressure has shaped them in the same way, leading to this extraordinary evolutionary convergence.

Africa and of which only the geographic "extremes" have been found? Future finds in the intermediate region of Africa (in which no fossils have yet come to light) may provide an answer.

The members of the group can now be seen crouched on the ground, digging for tubers—nutritious vegetables with a high water content. The *robustus* are also vegetarians. Some are chewing tough fruits with thin skins that they found along the way.

What else makes up their diet? Analyzing the signs of wear on their teeth under the microscope, Alan Walker, of Johns Hopkins University in Baltimore, has discovered that there are no traces of grass, bones, or roots. There is, however, an overall similarity to the way in which chimpanzees' teeth wear down, and they eat not only fruit but also termites, ants, sap, eggs, and small animals.

The tuber diggers seem to be holding something in their hands: perhaps a piece of bone, used as a shovel. Does that mean that *robustus* used tools in a rudimental way?

Two researchers, C. K. Brain of the Transvaal Museum and Randall Susman of the State University of New York at Stony Brook, who carefully examined numerous fossil bone splinters found with the remains of *Australopithecus robustus*, were led to believe that the furrows observed under the microscope are not accidental but the result of intentional use of the bone. Thus, like the ballistic analysis carried out by the police lab to determine whether two bullets were shot from the same gun (by comparing the lines left on the bullet), they took pieces of actual bone and went to dig tubers in the same region. They found that pieces of bone used as shovels and hoes to dig out tubers become furrowed in exactly the same way as the fossil bones.

That *robustus* were able to use tools is confirmed by reconstructions of their hands: they were similar to ours, with straight fingers and a large thumb, indicating strong muscles and a significant capacity for manipulation. The mobility and shape of their wrists were also very similar to ours.

Nevertheless, many believe that the remains of tools from that period (both bone and stone) did not belong to *robustus*, but to another very important actor who will be discussed in the next chapter: *Homo habilis*, a contemporary who lived in the same area (and also farther north in the area inhabited by *boisei*).

So, the corpus delicti probably did not belong to *robustus* but to another individual, much more intelligent and *habilis* (Latin for skilled), who was soon to come on the scene.

5

Toward Human Beings: *Homo habilis*

The First Species Named *Homo*

We have reached a crucial point in our story. For the first time individuals given the name *Homo* appeared on the scene.

Who were they?

Actually, the fossils do not provide a very accurate description. But various evidence indicates that they played a leading role in our evolution. Paleontologists are, in fact, agreed that they are our direct ancestors. Among the myriad of hominids that populated remote prehistory, *Homo habilis* is "definitely" a station we passed through.

But why *Homo* (man)? And why *habilis* (skilled)?

Actually, one has to be careful not to take the names and "labels" given to fossil remains too literally. They often do not reflect a rational classification and are based on personal convictions or are related to other finds. *Homo habilis* has actually been known by various names—seven to be exact. In this case, however, the current name seems to be the most appropriate one, in that these hominids had innovative features that reflect a transition toward human characteristics, above all, a larger brain and the ability to manufacture stone tools.

Where did they come from? This is still an open question. As in the case of human beings, parts of their genealogical tree are still missing. It is not quite clear who among the various hominids we have discussed so far were their ancestors.

Some people, like Donald Johanson, feel that *Homo habilis* is the grandchild of Lucy; that is, *Australopithecus afarensis* gave rise to the line from which we descended. Others, like Richard Leakey, believe that *Homo habilis* did not descend from Lucy but traveled on a parallel track, the previous stations of which have not yet been discovered.

As already mentioned, the real problem with the study of our origins is the scarcity of finds providing information to complete the puzzle. Then again, it must be stressed (and we will return to this shortly) that there was not a single kind of *Homo habilis* but probably a variety that were related, though very different from one another.

However, all were considerably more evolved than the Australopithecines; in particular, their skulls (and probably their brains) were much more developed. The volume of the brain rose from 400–500 cc (in Australopithecines) to 600–800 cc; that is, it increased by one half (with a similar body height of 45 to 50 inches). This significant evolution is reflected in the appearance of the first chipped tools.

Study of the skull of *Homo habilis* reveals evidence of a number of transformations approximating human characteristics. For example:

- the impressions of the blood vessels on the inner surface of the skull cap appear to be more ramified;

- the head is rounder;

- the associative areas are more developed;

- the brow is less pronounced;

- the foramen magnum (that is, the hole at the bottom of the skull through which the spinal cord passes) is further forward, giving the head a more erect and less ape-like position;

- the first signs of a chin are apparent;

- the teeth start to resemble ours, those of omnivores;

- the third molars (the famous wisdom teeth) are smaller, a typical human characteristic. The roots of the teeth are also smaller;

- the femur is indicative of the permanent upright position.

What evolutionary pressure led in this direction?

No one is able to answer this question today, but some hypotheses can be advanced. One key to the problem could be the way in which

19. "Gathering" meat from carcasses is an ancient strategy that may have begun with the *habilis*. Continued by *erectus*, it is still practiced by certain populations today. Vultures overhead indicate the presence and location of carrion.

food (a basic element in the survival and adaptation of a species) was procured.

Where will this line of investigation lead us?

From Vegetarians to Predators

Let's return to the African savannah about 2.2 million years ago (that is approximately the period to which the remains of the *habilis* date). The last Australopithecines, discussed in the previous chapter (*boisei* in the east and *robustus* in the south), were still walking the savannah. Thus, *habilis* was a contemporary.

These Australopithecines, as indicated in the last chapter, were living mills, well adapted to eating plant matter. The teeth of *habilis*, on the other hand, show a strong adaptation to frequent meat-eating. But in order to eat meat, you have to procure it.

Unlike plants, meat moves. It doesn't take much intelligence to find vegetables; but it does to hunt meat. It requires direct competition with the prey (hunting) or with another predator (stealing). It requires quick reflexes and the ability to improvise.

Furthermore, the energy obtained from the food must more than compensate for the energy consumed in capturing the prey; this equilibrium can be achieved only through intelligence. Thus, it takes strategy, organizational ability, and versatility to procure animal protein.

How did behavior evolve in the transition from these early vegetarian Australopithecines into predatory *habilis*?

Actually, all evidence indicates that the consumption of meat did not start with *habilis*; Australopithecines probably integrated their diets occasionally with meat, as chimpanzees do today. In fact, chimpanzees generally eat fruit, leaves, sprouts, and seeds; but now and again they capture and eat prey such as small mammals, birds, young baboons, and, in some cases, even human children.

It can, therefore, be hypothesized that with the emergence of the line that led to *habilis*, the occasional consumption of meat gradually became more habitual.

Looters of Carcasses

Actually, it was not necessary, at least in the beginning, to work out complex strategies for procuring meat. A very simple system for "capturing" prey already existed, and that was picking the remains of animals killed by other predators. Robert Blumenschine, of Rutgers University, has studied the sequence in which predators devour their prey: (1) the hindquarters, (2) the ribs and forequarters, (3) the head, (4) the marrow of the hind bones, and (5) the marrow of the fore bones.

There is one place on the savannah where prey can still be found in reasonably good condition," says Blumenschine, "and that is in the vegetation along the streams. Lions often attack here and abandon their prey after having eaten their fill."

Since hyenas were not frequent visitors of these places, large bones containing marrow, as well as left-over bits of meat and fat, were available for a few days (naturally suitable tools, such as "choppers," were needed in order to get at the marrow).

Habilis may have kept an eye out for vultures for this reason. Vultures circling overhead on the savannah almost always mean that food is nearby and that there will be little left for latecomers.

Hyenas also use this technique; they constantly scan the sky. And so do the Hadza, hunter-gatherers who live in northern Tanzania and systematically plunder the prey of lions, such as our ancestors once did. During the rainy season, they are very attentive to the flight of vultures and to the nocturnal cries of lions and hyenas. At the slightest signal, they drop whatever they are doing and run to the site of the kill, driving any intruders—including lions—away.

But isn't stealing the spoils from carnivores dangerous? This is an extremely interesting question. In the opinion of John Cavallo, from Rutgers University, who is doing some very innovative studies on leopards, *habilis* were able to draw from a natural pantry involuntarily made available by such dangerous carnivores as leopards.

As mentioned, after killing gazelles, leopards hoist them into the trees to keep them out of reach of hyenas and jackals. But once safely stored away, they do not devour them immediately; they sometimes leave them unguarded for hours or even days before eating them.

Cavallo feels that this made it possible for hominids to find fresh meat regularly in certain points of the savannah, without having to compete with hyenas or contend with dangerous carnivores.

The Keys to the Problem

Although for a long time hunting mainly meant hunting for carcasses, the emergence of other qualities gradually allowed for the development of predatory and other skills which made *habilis* a totally new figure in evolutionary history.

We don't know exactly how things went, but a number of factors must have exerted selective pressure toward more intelligent behavior. Would you like to ponder on this unsolved enigma yourselves? Below is a list of probable factors; try to find the most credible sequence.

a) the ability to procure meat (predation, looting, scavenging);

b) the use of tools (for example, to cut meat or break bones and extract the marrow);

c) the development of manual abilities for the manufacture and use of tools;

d) the organizational ability to hunt in groups (socialization, sharing of food);

e) language, for better communication and collaboration in the group;

f) the development of the volume of the brain and the nervous system.

Many combinations and sequences are possible. For example, d, f, and e could permit a, c, and b. Or vice versa. The sequence we prefer, however, is f, e, d, b, a, c, or *feedback*.

In other words, it is unlikely that any given sequence is better than the others: all factors were intertwined and interacted contemporaneously (feedback). They all affected the gradual evolution of the human condition and, above all, human behavior. And perhaps, for the first time, cultural behavior became a factor in selection.

This was the beginning of intelligence.

The Prehistory of Technology: Stone

So, looking back to that distant "moment" (which actually lasted hundreds of thousands of years), it's obvious that *Homo habilis* was the first being to possess the main quality underlying the success of our species: the ability to affect the environment.

In our world, know-how and technology are important, that is, knowledge and the tools needed to apply that knowledge. That is how we have overcome the constraints of our environment. Unlike animals, we do not have to wait for wings to grow in order to fly because be have airplanes. We do not require gills to go underwater because we have diving equipment, nor do we require claws to dig because we have bulldozers. We no longer depend on the laws of biological evolution: we have made our own rules. It is as though we could suddenly move our rooks diagonally or jump with our bishops in a game of chess—we would be sure to win. And that, in fact, is exactly what we have done.

As intelligence and tools developed, prehistoric beings started to influence their environment. The manufacture and skillful use of a stone tool allowed them to break open animal bones and have access to a new source of food: marrow. They no longer needed such strong jaws; nor were strong fingernails necessary for digging up roots and tubers. But above all, a larger brain with more refined and ramified nerve endings gave them a computer with which to solve problems in a creative way.

Homo habilis was probably the first being to be influenced by selective pressure based on mental rather than physical characteristics. Strength, speed, and perception were no longer the only criteria. Creativity, language, and behavior gradually started to dominate. These are the characteristics that determined success: they made it possible to procure food more efficiently; to hunt collectively; and later, to invent fire, to migrate, and to build more effective tools with which to hunt, fish, and build shelters.

Who Chipped the First Stone?

This process started with *Homo habilis*, although it had been developing for some time. In fact, the first manufactured stone tools date back

20. "... A closer look reveals dark shadows that are concentrated in small groups in the branches. Still cold, the creatures await the first rays of sunlight before moving. ..."

to a slightly earlier period (the most ancient *Homo habilis* known to us dates back to 2.2–2.3 million years ago, while the most ancient tool dates back to 2.3–2.5 million years). Of course, there may be uncertainties about the dating and interpretation of fossils (and, in fact, there are divergences), but on the whole it is quite clear that the appearance of manufactured tools coincides with the emergence of hominids with larger brains.

The difficulty in tracing the evolution that led to *Homo habilis* is also explained by the fact that almost no fossils have been found from the period ranging from 2 to 3 million years ago. This obviously makes the association between hominids and tools more problematic.

In the preceding chapters, an important "gap" between 5 and 9 million years ago was mentioned. This is another difficulty that prevents us from following the transition more closely; only the discovery of increasingly complete finds by field researchers can bridge this gap.

It is not surprising that stone tools may have been used by some pre-*habilis*, prior to the appearance of *habilis*. Tools made of perishable materials such as wood, bone, or quills and branches, were probably manufactured and used in even more remote times, only no traces remain. It is not unreasonable to think that evolution was gradual at this stage as well.

Nevertheless, only with *Homo habilis* do all these factors come to light and reach a "critical threshold," accelerating the transition from a selection based almost solely on biological attributes to one based increasingly on cultural criteria.

While selective pressure may have been based on acceleration for a gazelle, hearing for a hare, or sight for a falcon, from that time onward, selective pressure for hominids started to be based on language, memory, creative association, being able to imagine the future, the ability to analyze and synthesize, and so on. All these things, which are now permanently incorporated into the structure of our brains and allow us to define ourselves as *sapiens sapiens*, started to develop at that time.

A Pallet in the Trees

Thus, the appearance of *Homo habilis* approximately two million years ago is a fundamental landmark in evolution—a station at which the tracks were switched, sending the apes in the direction of human beings.

Therefore, it would be interesting to understand better who these people were, how they lived, how they procured their food, how they moved in their environment, and how they slept and interacted with others. The dream of every anthropologist is to be able to observe a group of *habilis* from some secret lookout for an entire day on the savannah. That is impossible, but why can't we at least try to imagine what it must have been like—even if it is only an approximation?

Imagine a group of *Homo habilis* waking up in the trees.

Why in the trees? It may have been the only safe place to sleep. Sleeping is dangerous: during sleep one is vulnerable to predators and the savannah is full of them—a night alone on the savannah is enough to convince anyone of that. Hyenas are only one kind of terrible nocturnal raider. Many a camper who has preferred to sleep under the stars rather than in a tent has been horribly mutilated.

If you had to spend a night on the savannah, what would you do? You might do exactly what all the primates do: climb a tree. And you might make yourself a small pallet of boughs, as the chimpanzees do. Trees are not hazard-free—leopards climb them (and, in some cases, so do lions)—but they reduce risks to a minimum.

We think it reasonable to assume that *Homo habilis*, who had not yet discovered fire, resorted to this solution. Although land dwellers, they probably took to the trees at night for shelter. In any case, all their ancestors almost certainly did so.

So, what was a typical day in the life of a small group of *Homo habilis* on the savannah of eastern Africa like two million years ago? The knowledge gained from the remains and "fragments" of fossils, paleoclimes, paleofauna, and so on, can help us use our imaginations more effectively.

A Day in the Life of the *habilis*

It is dawn. Cold, biting air shrouds the grey outline of a huge lake. The silence is broken now and again by the short, repeated shrieks of birds or isolated cries. In the distance, a group of Deinotheria—proboscidates with tusks oddly turned downward—slowly moves through some scrub. Not far away are gazelles, wart hogs, and hippopotami. Brightly colored birds flutter among the acacias and the euphorbias.

21. ". . . *There are also tubers and fruits in this area. The group disperses and the members start to pick and gather nuts and berries.* . . ."

A closer look reveals dark shadows that are concentrated in small groups in the branches. Still cold, the creatures await the first rays of sunlight before moving. Many are still dozing on pallets made of branches and leaves gathered in the area.

A chorus of guttural sounds—short messages from one individual and from one group to another—can be heard. Language is not yet articulate, but sounds are already coordinated.

Suddenly, some figures stretch to their full heights and, grasping the branches, start a slow and circumspect descent to the ground. We can finally see them in action: they look very familiar, quite different from any apes or chimpanzees.

Upon reaching the ground, the first hominids stand up and start to survey their surroundings: dawn is the ideal time for hunting lions. The others descend. They walk effortlessly in the upright position, as we do, and leave footprints that are identical to ours. It is obvious that they feel more at ease on the ground than in the trees. Some of them handle sticks, others pick up chipped stones left under the trees.

The sun is now a golden sphere on the horizon and the air is starting to warm. Walking through the underbrush toward the lake, they never move too far away from the trees, which could at any time constitute a safe refuge for these beings that do not have the claws, fangs, or the speed required to survive on the savannah. They frequently stop to investigate suspicious odors. Experience has taught them that the brush on the shore of the lake is a favorite hiding place of predators.

Once they have received the "all clear," the hominids spread out along the lakeshore to drink. Some bend over, touching their lips to the water, others use their hands as cups. This technique is often used by mothers with small children. The barely rippling waters reflect faces that are somewhat frightening: shaggy, perhaps curly hair (as protection from the sun) covers a low brow. The cheekbones are pronounced, the nostrils are flatter, the chin is receding. The face, like the rest of the body, is almost hairless. The skin is dark: another adaptation to the savannah sun. The dark eyes have a human expression, reflecting emotions similar to ours.

A few hominids nervously eye two crocodiles floating stilly in the water. Some walk along the shore in search of food or useful objects like stones or sticks. Curiosity is one of the outstanding characteristics of the *habilis*. One of them seems to be limping: arthritis in the foot, a rheumatic ailment, makes it difficult to move quickly. This may well prove fatal sooner or later.

The group leaves the shores of the lake and returns through berry-laden bushes. There are also tubers and fruits in this area. The group disperses and the members start to pick and gather nuts and berries with one hand, clutching their booty to their chests with the other.

This task is shared by all; in this way, the youngsters soon learn which berries are edible. Their infancy is incredibly long for an animal of the savannah. This allows them to learn much and develop new behaviors and strategies for survival. Their long infancy is one of the reasons for the success of these hominids, but prolonged child care limits the mother's freedom of movement and her efficiency in procuring food. In fact, after a brief vocal exchange, a group of males sets off in search of other sources of food. A few young males go along, eager for new experiences.

The sun is high in the sky. Skirting the tall grass, the group comes

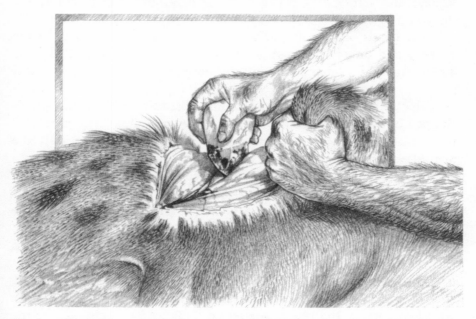

22. ". . . *Despite their strength, they cannot strip all the meat from the carcass with their hands and have to resort to their* choppers *(and sharp splinters from them) to cut the tendons. . . ."*

upon flocks of guinea fowl. The result is a foray for eggs and fowl, captured either through pursuit or simply by taking the animals by surprise. Some members of the group stop to dig with their pointed sticks. Tiny leaves on the surface betray the presence of edible tubers.

But the favorite activity of these hominids is scouting out places where the huge herds of gazelles and antelopes have spent the previous night or day. There is always the remote possibility that they might come upon some isolated animal—injured, sick, or trapped. Unable to hunt large prey themselves, the hominids take advantage of those who can: the carnivores. The object of their quest is a carcass or, better yet, freshly killed prey.

They finally find one, but the pickings are meager. Undaunted, the hominids start to break open the long bones and extract the marrow, using stones that they gathered the day before near a large outcropping of quartzite. They often go there and other places to gather blocks of quartzite or basalt for the manufacture of tools. The technique is simple: find a rock that fits well in the hand and chip it to create a sharp cutting edge.

The hands of these hominids are still primitive. The bones betray a strong and rather imprecise grasp as brain development is still in its initial stages (and innervation and the processing of movements is still poor).

That is why tools are also primitive: basically pebbles and stones chipped or flaked in a rudimental way. They are called "choppers" and will remain unchanged for hundreds of thousands of years; only the mature *erectus* will start to create new and more sophisticated tools.

Some of the *habilis* have discovered another carcass. It is more recent and some hyenas and jackals are tearing it to pieces. This is a real stroke of luck. Shouting and brandishing their sticks, they throw stones to drive away the animals, before eagerly pouncing on the meat. Despite their strength, they cannot strip all the meat from the carcass with their hands and have to resort to their choppers (and some splinters) to cut the tendons.

The meat is cut into pieces and partially eaten. The fleshier pieces are taken back to the rest of the group. Followed at a distance by the hyenas, the hominids head back carrying meat, bones, tubers, and eggs. They were very lucky . . . this time.

While returning, a young hominid sets off through the tall grass

23. ". . . There is a moment of tension; some individuals pick up stones and sticks. But neither hominid takes a step toward the other; neither invades the narrow strip dividing the two worlds. . . ."

in pursuit of a guinea fowl. Suddenly he stops dead in his tracks: standing before him, at a distance of less than thirty feet, is another hominid that is staring intensely at him. But the other hominid is quite different. Taller and stockier, he has a much broader face. Behind him, others like him, including females and children, can be seen. This is an *Australopithecus boisei*, a contemporary of *Homo habilis* and an extinct branch of our evolutionary tree.

There is a moment of tension; some individuals pick up stones and sticks. But neither hominid takes a step toward the other; neither invades the narrow strip dividing the two worlds. The *boisei* are vegetarians, walking millstones, as mentioned previously, with enormous teeth to grind tough plant matter. This *boisei* is probably more frightened than the young *habilis* because he knows that the other is a meat-eater and could go after his young if circumstances permitted it. The two groups often meet on the savannah, but they are neither enemies nor rivals because they occupy different ecologic niches. The *boisei* see the *habilis* as opportunistic omnivores that are not to be trusted, while the *habilis* see the *boisei* as peaceful herbivores that are too large and well organized to be attacked.

Only in the late afternoon do they reach the clearing and join the rest of the group, which is very excited at the sight of such good hunting. Cries and shouts are mixed with guttural messages. The food brought home (along with other fruits and tubers gathered in the clearing in the meantime) is shared by the various family groups; such cooperation is unusual for primates and one of the reasons for the success of these hominids.

The sun is setting, giving a red tinge to the sky and the smoky spiral rising from the nearby volcano. A light breeze has come up, the air is cool. The group walks slowly toward the trees where their pallets from the night before are located; they will probably move on in a few days in search of new sources of food and new trees in which to sleep.

From their perch in the trees, they watch the stars come out. The guttural sounds accompanying all the activities of this ancient human being gradually subside.

Another day almost two million years ago has drawn to an end.

Evolution in Time and Space

This imaginative reconstruction based on the facts currently available has given us insight into the daily activities and problems of our distant ancestors. Although it is known that these hominids and their environment led to the real evolution toward modern human beings, the route is not completely clear.

As mentioned at the beginning of this chapter, there were several quite different forms of *Homo habilis,* not only because of the "vertical" evolution in the 600–700 thousand years in which they probably lived (the first *habilis* were similar to the archaic Australopithecines, while the later forms were similar to early forms of *erectus*—their descendants), but also because of the "horizontal" evolution in that time, that is, a series of ramifications in the same period that probably led to various kinds of *habilis.*

All evolution on the earth has demonstrated that development is different in different ecological niches (just think of the human race and how different the Pygmies are from the Watussi, who live in the same region).

In other words, evolution must not be seen as a column of soldiers lined up in single file, but more as an army scattered randomly over the terrain. Some divisions advance, others halt or surrender, or set out on reconnaisance missions and end up finding shelter in distant places. In the end, only a few divisions survive, handing the banner on to new soldiers (their children) who will continue the march.

That is why the fossil remains indicate such differences in *habilis.* Furthermore, like *Australopithecus,* remains of *habilis* have been found in various parts of the African continent.

There is a strange synchrony between the appearance of these forms of hominids in eastern and southern Africa: frail Australopithecines lived in these two places approximately three million years ago; then Australopithecines of the stockier type appeared in both areas about two million years ago, and only a short time later, *Homo habilis* came on the scene.

Does that mean that these people migrated? Or did evolution take place (at least in some cases) along separate but similar lines? It is interesting that similar forms of *habilis* appeared almost contemporaneously in very distant places.

In other words, *Homo habilis* did not exist in eastern Africa alone.
Remains of *habilis* dating back between 1.5 and 2 million years have
also been found in South Africa, in particular, at Sterkfontein and
Swartkrans. But the most interesting specimens have been found in eastern Africa. A closer look may show us some of the differences between
them. Like prison inmates, the individuals are identified by numbers:
1470, 1813, and OH62.

Two Classics: 1470 and 1813

These numbers refer to the progressive classification of finds. The first
two were found in Kenya near Lake Turkana in 1972 and 1973, respectively. The third was found in Olduvai, Tanzania, in 1986. (OH
means Olduvai Hominid and 62 is the progressive number given to
the remains of hominids found in that area, whether fragments or a
single tooth. In fact, only a single tooth of the next hominid, OH63,
has been found.)

It only takes a glance at the first two—two skulls (1470 and 1813)
—to realize that they are quite different: the cranial volume of the first,
for example, is calculated to be approximately 750 cc (the volume of
our brain is 1450 cc on the average), while the other has a volume
of only 510 cc, although it lived at the same time or even later (that
is, 1.7 million years ago rather than 1.9 million years). How can this
important difference be explained?

First, the individuals were quite different in stature and size; some
experts think that they are male and female, claiming that sexual dimorphism (males being larger and females smaller) was still very accentuated at that stage in evolution.

Others feel that the difference has nothing to do with gender. A
look at the shape of the skulls (or the dentition) is enough to see that
the individuals were quite different: although more recent, 1813 has
a more receding brow and a greater resemblance to more ancient hominids. The differences are so great that some people maintain that they
belong to different species.

But what does detailed study of the two skulls reveal, aside from
these apparent external differences?

Number 1470 (which was patiently reconstructed from 320

fragments) was the subject of debate among its finders from the very beginning. Richard Leakey considered the hominid the true ancient ancestor of human beings, a line parallel to the Australopithecines (but at that time dating was incorrect: the skull was thought to be 2.9 million years old rather than 1.9 as discovered later). Alan Walker thought the skull belonged to an *Australopithecus,* who just had a larger brain.

Yoel Rak, an Israeli expert in the facial structure of hominids, points to a less apparent characteristic of 1470: the structure of the face (rather concave) and certain features linked to mastication resemble those of the robust forms of Australopithecines. This could be a convergence brought about independently by similar chewing habits.

Number 1813, on the other hand, does not display these characteristics; its teeth are also quite different and much more similar to ours.

The Surprising OH62

But the difference is even more evident in OH62, found in Olduvai Gorge in 1986 by Tim White and Donald Johanson. The expedition was organized in collaboration with the Centro Studi e Ricerche Ligabue of Venice. The find dates back to more or less the same time (1.8 million years ago), but it has quite a different structure.

During the excavation, in which one of the authors of this volume participated, more bones were found of the body than of the skull and, thus, for the first time a hominid from that period could be examined in its entirety. While the fragments of the skull are insufficient to calculate the volume of the brain, the bones of the body proved to be a real surprise.

The individual is an unusually small female (about three feet tall), but not a child: she has adult teeth that are quite worn down, suggesting an age of approximately thirty—very old for that time. The most interesting feature, however, is the length of her arms: her hands practically reached her knees, giving her a rather ape-like appearance.

The relationship between the length of the arms and the legs can be shown clearly by comparing the humerus and the femur. In the chimpanzee, the humerus is longer than the femur. In Lucy, the proportions are reversed: her humerus is only 85 percent of her femur (in modern woman it is 75 percent). OH62 is almost like a chimpanzee: at least

95 percent. Many interpretations can be given for this, but the investigators who found this specimen believe that it was an adaptation to arboreal life.

In other words, contrary to what one might expect from beings that belong to a more advanced evolutionary line capable of living a more independent life on the ground (thanks to the use of tools and the development of intelligence), there were at that time varieties of *habilis* that had adapted or readapted to life in the trees.

Does that mean that everything has to be revised?

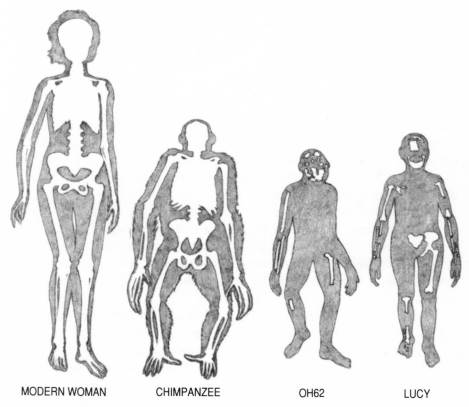

MODERN WOMAN   CHIMPANZEE   OH62   LUCY

24. Hominid OH62, a woman approximately three feet tall, had surprisingly long arms that almost reached down to her knees. She lived 1.7 million years after Lucy, but had a humerus/femur ratio close to that of the chimpanzee: 95 percent (in the chimpanzee, the two bones are almost the same length). The ratio in Lucy is 85 percent; in modern woman, it is 75 percent.

The problem is that in classifying the so-called *habilis*, very different individuals have been grouped together (this also results from the difficulty in "reading" the finds; if entire skeletons were always available, everything would be much easier).

Fewer problems exist for other types of hominids. For example, the remains found in different areas and even in different countries are substantially the same for the robust kind of Australopithecines (in particular, for the famous "nutcracker") which lived at the same time as *habilis*. This is also true for *afarensis*.

For *habilis*, though, the variety of shapes suggests different groups (or populations). Or even different species. But in the absence of concrete information, no one is really in a position to sketch a picture of these hominids.

The Tracks of Evolution: Toward *erectus*

Summing up, various kinds of hominids lived in Africa in that period: on the one hand, the robust forms of Australopithecines (the "nutcracker" *boisei* in the east and the *robustus* in the south, both bound for extinction) and on the other, probably various types of *Homo habilis*, only one of which, however, was destined to give birth to the line that would lead to modern human beings.

Which one it was cannot be established at this point because there is another gap in fossil findings between *habilis* and *erectus*. There are some signs of the tracks and a couple of ties but no distinct line to speak of. This is because the railway network of evolution, as already mentioned, is extremely ramified and extends in many directions: most of these lines lead toward unknown destinations or gradually turn into dead ends.

But that always happens: in evolution parallel tracks are everywhere. For example, we can still see (although in different forms) the various evolutionary stages we passed through from the beginning of life on earth: bacteria, fish, reptiles, mammals, and primates. They are the "vertical" descendants of those forms which were only transition stages for us. These primitive forms are still in existence today, giving us a horizontal view of evolution.

The same kind of horizontal display can be reconstructed for the period that dates back to 1.6–1.8 million years ago. At that time, beings representing different evolutionary lines could be seen (and encountered) in Africa: the last of the Australopithecines; various kinds of *habilis*; and another emerging hominid, taking over from an as yet unidentified *habilis* variety in order to continue the journey toward human beings —*erectus*.

Much attention will be given to *erectus* because it is a central character in the evolution of the human species.

Naturally, the name should not be misleading; as has already been shown, the erect walking position appeared at least two million years earlier (3.7 million years ago) with the Australopithecines. But Eugene Dubois did not know that. Dubois was the Dutch physician who discovered the femur and other pieces of this hominid in Java in 1892 and thought he had found the first of our ancestors that was able to walk on two legs—the missing link between humans and apes. That is why he called it *Pithecanthropus erectus*, which later become *Homo erectus*.

Thus, *erectus* is an all-important character in our story because it was the uncontested star of the show for over a million years—and not only in Africa, but in Europe and Asia as well. *Erectus* marked the beginning of migrations outside of Africa, the first adventures in territories quite different from those of origin, and the evolution toward beings that were increasingly similar to us.

The relationship between *erectus* and *habilis* (at least those found to date) is not clear; we do know, however, that an individual with very special characteristics already lived 1.6 million years ago.

Further study requires another journey in time back to the east African savannah and the shores of Lake Turkana, in present-day Kenya.

6

Homo erectus: The Turning Point

The Young Giant

Horses and hippopotami graze on the scrub-covered plains surrounding
a shallow, placid lake. The body of a young male lies in its muddy
bottom. The cause of death is unknown. Was he attacked by an animal?
This does not seem likely as the body bears no traces of violence. His
death remains a mystery.

The boy is very young (his teeth are not yet those of an adult),
but he is surprisingly tall: a bit over five feet. He would be one of
the tallest in a class of sixth-grade children today. If he had reached
adulthood, he would have been about five and one-half feet tall.

Sinking into the mud at the bottom of the lake 1.6 million years
ago, the young *erectus* was gradually covered by a fine layer of brown
silt. Slowly the skeleton fossilized, remaining perfectly intact to the present
day. This is the most complete body ever found from that time: only
the left arm, the right forearm, and the greater portion of the feet are
missing (and the search goes on for the remaining pieces).

"As soon as we found it," says Alan Walker, "we realized that
it was unusually tall for a twelve-year-old and thought it was an abnormal
individual. Then, carefully studying a score of femurs of *erectus* at the
Nairobi Museum, we realized that they were all long. No one had ever
noticed it before, because only fragments had been found. Arbitrarily,
without any proof, it had been thought that the *erectus* were not tall."

The laboratory revealed other surprises: the young man's chest has
twelve ribs, like ours, but he has six floating ribs (we have two), making

101

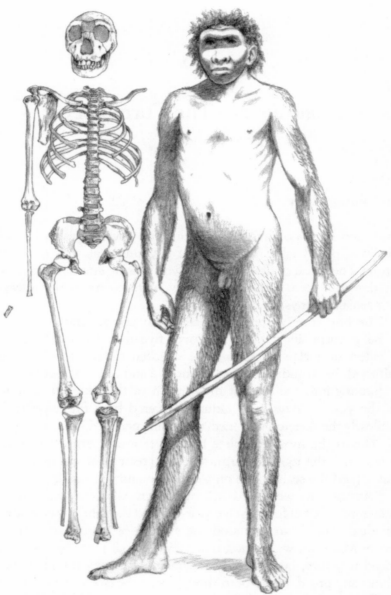

25. The most complete skeleton from the past: that of a young *erectus*, around twelve years old and over five feet tall. He died near Lake Turkana 1.6 million years ago. Sinking into the mud at the bottom of the lake, his body was fossilized and now provides invaluable information on the most ancient *erectus*.

his trunk much more flexible. His spine is also different; the vertebral processes are oriented differently and the vertebral canal, through which the nerve endings pass, is only half the size of ours. In fact, Richard Leakey, whose team discovered the skeleton, named KNMWT 15,000, said that in this hominid "practically every piece of bone shows minute but unquestionable differences from modern man."

But the overall structure is similar. And one important feature makes us feel very close to him: if we look at the impression left by the brain on the skull, the area of Broca—the part of the brain in modern human beings that controls language—is evident. This young man probably used language to communicate with his fellow hominids (even though the matter is slightly more complex, as will be seen in the next chapter).

The only indications of the intelligence of KNMWT 15,000 can be drawn indirectly from the volume of the brain. As is to be expected, it was not highly developed: measuring less than 900 cc (perhaps only 700–800). This is small compared to the size of the brain of later *erectus* (the volume of which varied from 800 to 1225 cc for a height of five feet). However, no definitive conclusions can be drawn from simple measurements like these; as those who discovered KNMWT 15,000 maintain, brain development in this case was probably not complete since the hominid was so young.

The Bust of *erectus*

Study of this early *erectus* may lead to many other discoveries in the future. This individual seems to be a starting point for that long evolutionary journey which, in the course of a million years, led the hominids out of Africa and triggered a series of extraordinary events.

A million years is a long time: equal to approximately 50 thousand generations. In that time, starting from a basic model, *erectus* went through a series of changes in time and space that generated a small universe of groups and individuals.

Many *erectus* (mostly just fragments) have been found throughout Africa, Europe, and Asia. All this material has been used to piece together a portrait of this ancestor.

If an investigator were to commission a bust of an *erectus*, like those

26. The skull of the *erectus* from Peking, with a reconstruction of the muscles and the face based on the information available today. This is an attempt to bring us into closer contact with a reality that will always elude us.

of Roman emperors or philosophers, from a sculptor, providing all available information, what would it look like?

The first thing the sculptor would have to do in modeling the face out of clay is to make a heavy brow (like boxers after a tough fight): this is the so-called "visor," typical of gorillas. The forehead would then have to be flattened out until it almost disappeared. Finally, the sculptor would have to apply pressure at the temples to create depressions between the eyes and the ears and fill them with the strong muscles of mastication (seen from the top, it looks as if the skull is wearing a diving mask). The mandible takes a lot of clay. It has to be massive, like the shears of a trap. The chin is swept back with a clean diagonal stroke.

The eyes are set far apart under the "balcony" created by the brows. The nose is long and flat (but this would be left up to the imagination of the sculptor, because no fossil remains of the nose have ever been found). The lips would also have to be invented, but suited to the large teeth and the marked prognathism (the protruding jaw).

The final touch is the way the head is set on the shoulders: the presence of large bony protrusions at the back of the skull would suggest strong muscles between the neck and the spine. An *erectus* would probably look as if he had no neck—only a bundle of muscles directly connecting the head to the spine.

Placed beside the bust of Marie Antoinette, this bust would certainly cause experts in heraldry some difficulty (he was definitely an ancestor of Marie Antoinette, even though his neck would not fit under the guillotine!).

But what would we think of an *erectus* if he were alive today? Some people believe that, dressed in a white shirt and tie and, above all, a hat, an *erectus* could get by unobserved on the bus—especially in the evening. Others feel that he would soon be hustled off in an ambulance and institutionalized.

This would naturally be a mistake, because this *Homo* discovered fire and traveled as far as China; he was both Prometheus and Marco Polo. But the fact that he would be taken to an institution is significant in itself: Australopithecines and many *habilis* would have ended up in the zoo.

The transition from zoo to institution is, after all, the transition from animal to man. *Homo erectus* is an individual who strongly resembles us and who is clearly part of our ancestry, however remote.

Diffusion Outside of Africa

We're just joking, of course, and affectionately at that. With all the things they knew how to do (and did), *erectus* were instrumental in the transition toward increasingly intelligent societies and, ultimately, toward human society as we know it today.

Their developing intelligence and the increasing mastery of fire permitted *erectus* to start large-scale migrations from Africa to the Middle East, India, China, Indonesia, and, of course, throughout Europe.

The first remains of *erectus* were found by paleontologists in the nineteenth century. Different names were given to the species living in different places: *Atlanthropus* in northern Africa, *Sinanthropus* in China, and *Pithecanthropus* in Indonesia. We now know that they all belong to the same line, which spread out (over a long period of time) and probably gave rise to a variety of local subevolutions or races. Some people feel that the line actually produced different species.

Excavations in Java have led to the discovery of a wide range of individuals. These findings have caused much confusion because dating of the strata is impossible. It is as though the remains of skulls were picked up in the middle of the road: no one can tell what period they belong to. There is a certain succession of shapes, though: the oldest has a brain volume of 755 cc, while that of the most evolved is 1255 cc.

These finds (all considered remains of *erectus*) have been given various names: *Meganthropus*, *Pithecanthropus*, Solo Man, and so on. The last is considered an evolved *erectus*, perhaps even a primitive *sapiens*.

But what do we know about the life, behavior, and feelings of *erectus*? How did they hunt? build shelters? ensure survival for their children and their group?

Our reconstruction is set 400,000 years ago in a place much frequented by *erectus*: Zhoukoudian, in China (formerly Choukout'ien), a hilly area not far from Peking (Beijing). As always, the reconstruction is based on evidence and remains gathered by paleontologists over the years.

The Peking Man was a classic *Homo erectus*, and is often taken as a model for many of his physical and behavioral characteristics. Many exquisite skulls and traces of the way in which these people lived (the remains of hearths, tools, bones of hunted animals, etc.) have been found in the cave (or rather what is left of the cave after an ancient collapse) and in the Zhoukoudian area.

Despite constant, gradual evolution, the general physical structure of *erectus* remained the same for almost one million years in Africa, Asia, and Europe. Thus, the people we will soon be seeing in action are probably quite similar to those who lived on the African savannah or the European steppes at that time.

What follows is a description (reconstructed from evidence found in the field) that an eyewitness to events would recount.

Four Hundred Thousand Years Ago: Home from the Hunt

The landscape is familiar. The most characteristic feature is, perhaps, the crisp mountain breeze filled with the fragrance of spring. Elm and cedar forests cover most of the hills. Grasslands covered in blossom— mainly lilies—extend from the forests and skirt the numerous lakes and rivers, down to the valleys and the plains.

The forest is dotted with white spots, which could be karstic out-croppings of rock or patches of snow. Winters are cold and there is still some snow on the ground (especially in the higher regions) in spring. As soon as summer comes, though, the climate changes radically, becoming torrid.

The silence is interrupted by the sound of branches breaking underfoot in a thicket. Two individuals emerge from the brush, followed by a group of six or seven. Their stocky bodies are covered in skins and furs. They are not very tall, but from the ease with which they drag and carry the prey (two large deer, similar to the sika found in the Far East today) they have killed, it is obvious that they are very strong. The tip of a spear is still embedded in the side of one of the animals.

The group slowly follows the first two, who carefully look around for predators on the prowl. One of the two *erectus* is carrying a large beaver captured shortly beforehand in a stream; its fur will be useful next winter. Others are carrying berries and fruits: delicacies to be shared with the rest of the tribe.

It is not long since their return to this area (not far from present-day Peking) this year; they have been coming here for generations. This was the first hunt and it was definitely successful. As all *erectus* in Africa, Europe, and the Middle East, these ancient hunters were very skillful

27. "... *But after only a few feet, the oldest hunter stops and with a wave of his hand orders the others to the ground . . . : a saber-toothed tiger. . . .*"

in capturing all kinds of animals. But one animal is their favorite for its meat, its antlers, its bones, and, above all, its fur—deer. Deer are very abundant here. That may be the main reason that the hunters return every year.

After a long trek, the hunters decide to stop for a rest, giving the observer the opportunity to take a closer look.

The group is seated. Their strong bulging muscles can be seen through the openings in their furs. Their hands are large and rough, capable of a firm grasp, their legs short and muscular, used to long marches and long pursuits which usually end with the collapse of much faster but less resistant prey.

Their faces are very primitive: their brows protrude like a visor from a low and receding forehead. Below them are deep-set eyes.

They cannot be considered attractive in the current sense of the word; on the contrary, they look rather like animals. But their eyes have a different glimmer, a human look, which distinguishes them from all other beings. Intelligence beams from under the primitive and dark mask of the primate.

The group is now resting near a stream. Some individuals are drinking, cupping their hands. Other rest in silence, exhausted. Two are crouched to one side and exchange guttural messages in a rudimental but efficient language: the older of the two is concerned about the well-being of the younger, wounded during the hunt. A moment's distraction made him trip and hit his head hard against a rock. For a moment, the gush of blood from the wound made his companions think that he would not survive, but the young male now seems to be all right. A healthy individual, he will probably make it (as many others have, even in case of repeated fracture of the skull, which is very frequent). The resistance of these individuals, the product of millennia of natural selection, is extraordinary.

Another young member of the group is also on his first hunt. We can observe him as he passes. He can't take his eyes off some dark figures grazing less than a mile away, close to a herd of horses. They are bison, a much valued prey, which, however, require a special and very dangerous kind of hunt. Only the strongest and most experienced hunters are able to bring down an animal of that kind.

Furthermore, they require special spears that are thicker and heavier than the one this young male possesses. Above all, it takes a lot of strength, accuracy, and courage to spear such large beasts, which react violently to being isolated from the herd.

The *erectus* hunters set off again. They reach a river just as the sun starts to sink in the sky. Fording the icy waters with the two large catches and a wounded companion is difficult. A number of attempts have to be made before all are able to reach the other bank; they then continue toward the camp. But after only a few feet, the oldest hunter stops and with a wave of his hand orders the others to the ground.

Tension fills the air. One word from the elder is enough to make the blood of the entire group run cold: a saber-toothed tiger.

Their hands sweat as they clutch their spears with fire-hardened points. They are in a vulnerable position: out in the open with their bloody prey.

The young man's heart skips a beat as he catches a glimpse through the trees of the large powerful beast majestically striding across the clearing. Suddenly the saber-toothed tiger stops, raises its snout, and sniffs the air. It has picked up the scent of the freshly killed deer. The horrible white fangs descending on either side of its mouth are now quite evident. One bite would be enough to sever the head or the leg or slash the chest of the sturdiest hunter. Its long sharp claws can tear open even the tough hide of an elephant.

The tiger stands motionless, its short tail nervously flicking the air. Hours seem to go by as it observes the riverbank. Suddenly a gust of wind disperses the faint odor of the deer. With a final hesitant glance, the huge predator stalks away and disappears.

Camp is not far away and the very thought dispels the weariness of over two days of walking and hunting. The last leg—a climb through a forest of elms and cedars—is tiresome. The air has cooled down. During their last stop to rest, some members pick berries from the shrubs.

No one pays any attention to a skull lying on the ground not far away: it is a human skull, the remains of another *erectus* who was killed by a bear a few months earlier. Bears do not usually eat meat, but it may have come out of hibernation and been hungry; or it may have felt threatened and attacked. It had certainly scared the victim's two companions, who barely managed to escape to tell the story.

The hyenas and some "cuon" (a kind of wild dog) had probably mutilated the cadaver, tearing away pieces of meat and carrying them to their dens—perhaps a cave—to eat undisturbed (the bases of some of the skulls that have been discovered are smashed, as if attempts had been made to get at the soft brain).

No one came to bury this hunter. After death, individuals were abandoned (this still occurs in some primitive societies).

One of the hunters bends down and points to some tracks in the soft ground along the stream: they were left by a leopard. The group at the camp has to be warned.

The sun is low by the time they reach the camp. The first sign of the camp in the distance is a column of smoke rising through the trees. Then voices and shouting are heard. Some women holding bundles of dry sticks appear. They are at the foot of what the Chinese now call "Dragon Bone Hill," where the Zhoukoudian cave is located. The temperature has dropped further and patches of snow are still lying here and there on the ground. But it doesn't matter; excitement prevails over weariness and cold. Embraces, slapping of the hands, special kinds of touching and caressing—a complicated series of gesticulations based on a precise hierarchy and kinship—are exchanged in a burst of guttural sounds, gurgles, shouts, monosyllables, and simple words. Soon others emerge to help the hunters carry the prey into the cave.

The cave is huge: over 420 feet long and up to 120 feet wide in some places. The entrance faces east, toward the rising sun. The hunters are met with a familiar scene: children are playing close to a large fire. Nearby, some women are preparing berries while others withdraw to nurse their infants. Some fur pallets can be made out in the dim light. Bones, the leftovers from meals, are piled here and there. It looks cozy.

Fire is very important for this tribe of *Homo erectus*: it has permitted them to endure the harsh climate and to keep wild beasts at bay. (The carnivores have learned to avoid these human beings; only rarely do they manage to catch them unaware or undefended, and they have frequently felt the pain of fire or the sharp point of a spear.)

Sitting next to a pile of wood, an elderly member of the group is trying to soften the fur of a kind of sheep killed some time before.

Those who stay at home—the women, the children, and the few old folk—not only have to see to domestic duties, such as gathering berries and wood, but also have to tend the fire, stoking it and blowing on the coals now and again to rekindle the flame. Another important advantage offered by fire is in cooking: meat can be made more tender for those who cannot chew well, particularly children and the elderly.

The tribe seems to be made up of about thirty individuals. One hunter stayed at the camp because of a bad sprain. Surrounded by a mound of chips, he is sitting in a corner together with a young man, perhaps his son, fashioning stone tools. Some children gather around to watch.

Analysis of the implements left by the Peking people in the cave shows a clear evolution in the manufacture of tools. Improvements in technique led from large, rough tools to smaller, increasingly sophisticated, and more effective models. This was made possible by the advantages of social life: the refinement of traditions handed down from one generation to another.

The hunter is now crafting tools that are very light and no more than two inches long. They serve a wide range of purposes, from cutting to scraping. With one hand, he holds a piece of quartz in a vertical position on a flat stone. With the other, he hits it violently with vertical blows, sending chips flying from both surfaces. This technique, commonly used at that time in Zhoukoudian and known as "bipolar percussion," denoted a certain intelligence and skill on the part of the craftsman using it. In addition to accuracy and craft, the manufacture of good tools called for good materials. The stone which this hunter is working was taken from distant areas to which special "expeditions" were made for that purpose.

The two deer have now been dragged into a corner of the cave and expert hands have started to skin it with the help of sharp instruments. The procedure follows precise rules: no piece of fur must be wasted or ruined. Nothing is thrown away: bones, antlers, meat, tendons, or guts. Everything is either used or eaten, accompanied with berries, seeds (like those of the Chinese nettle tree), nuts, or fruits. There may even have been some simple characteristic meals that we will never know about. Fire certainly played its part in the cuisine of the "Dragon Bone Hill" tribe, as some half-burned remains demonstrate.

The sun has set and the first stars twinkle in the dusk. A thick mist engulfs the valley as the cold becomes more penetrating. The sliver of a moon faintly illuminates the gently wooded countryside, the hills, the plains, and the rivers, giving the remaining patches of snow a silvery blue sheen. A slight breeze gently nudges the tips of the pines. A wolf is howling in the distance.

Inside the cave, shadows of the ancient hunters seated around the fire are projected against the wall by the flames. The members of the group are occupied in various ways: eating meat or berries, playing, caring for the children or the wounded man, describing the hunt to others, or sharpening spears and tools. Some are simply gazing into the fire as if hypnotized. Although their faces are still different from ours, they have a much more human look in their eyes. They are like sculptures in the making, "unfinished" works.

The stars are now out in full force. In the semidarkness, figures huddle under the warmth of the deer furs. A day 400,000 years ago is drawing to an end.

28. ". . . *The stars are now out in full force. In the semidarkness, figures huddle under the warmth of the deer furs. A day 400,000 years ago is drawing to an end. . . .*"

How to Migrate

Of course, the scene just described has never been filmed or reported by a journalist, but there is much evidence to tell the story today: fossil pollens, bones, traces of carbon, and tools tell us about the plant life, fauna, fire, and human intelligence.

Zhoukoudian is only one of the many sites in which *Homo erectus* settled and lived. Starting from Africa, these early prehistoric travelers migrated (but it would be more correct to say spread out) all over the globe—across savannahs, steppes, drylands, forests, rivers, and mountains.

It is not known what made these hominids, who should have been interested in staying in the safe, familiar places of their ancestors, stray so far. What is certain, though, is that *erectus* were not explorers like Christopher Columbus. Their travels were not for adventure or conquest; they were probably not even spurred by the curiosity to see new places. Their travels were much more simply the sum of minor moves dictated by chance, the search for food, or the pursuit of animals, which probably moved as changes in climate opened up new environments. These moves were made over countless generations.

But why did they occur with *erectus* and not with other hominids?

Because *Homo erectus* opened up a new dimension. Intelligence led to autonomy. These creatures depended less on the environment, knew how to hunt and organize better, had more versatile tools, and could communicate experience.

It must be kept in mind that the time frame is extremely long. It is true that 8,700 miles separate Lake Turkana from Lantian (the most ancient site of *erectus* in China, much older than Zhoukoudian), but even at an average of 300 feet per day, the distance can be covered in less than 500 years. And there are 800,000 years separating the young *erectus* of Lake Turkana (that lived 1.6 million years ago) from the skulls found in Lantian (which date back 800,000 years); in theory the distance could have been covered more than a thousand times. Although these are only theoretical calculations (migrations did not occur in this way), they prove that distance is not a problem over long periods of time.

The same is true for animals: the planet is full of animals which migrated to its most distant and inaccessible parts. The search for food, the emergence of new ecological niches, and climatic changes gradually drove animals to spread out and diversify.

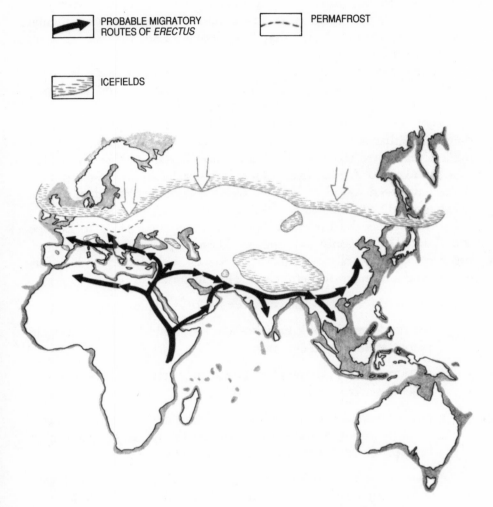

PROBABLE MIGRATORY
ROUTES OF *ERECTUS*

PERMAFROST

ICEFIELDS

29. One migratory hypothesis: the fall of the sea level as a result of the Gunz glaciation created natural land bridges for *erectus*. Skirting deserts, mountains, and areas of permafrost, they wandered where rivers, climates, and animals led them. Paleogeography may be able to make an important contribution in the search for their traces.

Italy, for example, was inhabited by rhinoceroses, monkeys, hyenas, and hippopotami. How did they get there? Certainly not by boat! If macaques were able to reach the peninsula over 2.5 million years ago, why shouldn't hominids have made it, climate and environmental resources permitting? And it may have actually been that way, but no proof has yet been found.

Hominids were, however, only one of the thousands of species living in Africa and not all migrated. For example, gazelles and antelopes, chimpanzees and gorillas did not leave the continent.

Therefore, the migrations could well have started with *erectus*—but then again, that is the only evidence we have. (Tools dating back 1.3 million years were recently found in Galilee, Israel, but that area is still closely linked to Africa. Even older, but unreliable and controversial traces have been found in Europe, like the choppers found in France which, according to their discoverers, date back 2–2.5 million years.)

What routes did the *erectus* follow on their migrations to Europe and Asia? We can only hypothesize. These hypotheses are, however, based on concrete factors, which provide us with considerable information and a number of useful clues. They include:

- the need for water (and therefore for rivers, lakes, or springs);

- the need for a suitable environment for survival (with compatible plants, animals, and climate);

- detours posed by geographic and natural barriers (bodies of water, deserts, glaciers, and mountains);

- environmental variations that could facilitate moves and migrations (lowering of the sea level during the ice ages, natural land bridges, climatic changes creating new habitats, etc.).

A map is useful for hypothesizing. Some possible routes followed by *erectus* on their way from Africa to the Middle East, India, Indonesia, China, and Europe are indicated on the map on the previous page.

Some traces of migrations can still be found along those routes (in the points that were ecologically most favorable at the time). But systematic research is lacking. Nor is there the right "exposure" (not to mention the problems of transformations in the terrain after thousands of years of civilization, traffic, and agriculture).

ICEFIELDS

NEW CONQUESTS
BY *ERECTUS*

30. Thanks to the retreat of the icefields, *erectus* could move northward in Europe and in Asia during the interglacial period.

The Possible Role of Glaciations

Although many fossils and sites have gradually been worn away by time, it has been possible to find a few. A striking fact is that the most reliable dates for the most ancient sites (from China to Italy, from India to Morocco, from Czechoslovakia to Russia, to France) all correspond to

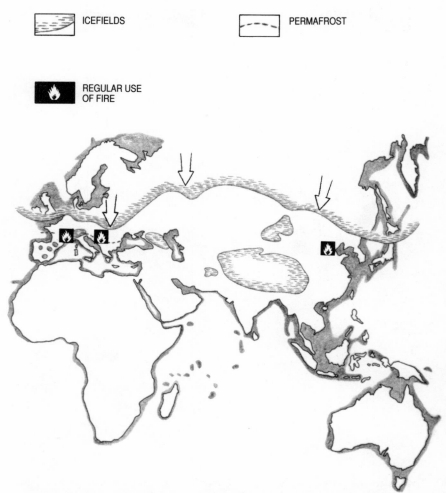

ICEFIELDS

PERMAFROST

REGULAR USE OF FIRE

31. The Mindel Glacial Age (300,000 to 500,000 years ago) caused new climatic changes: temperatures dropped considerably in the Northern Hemisphere to which *erectus* had migrated. The first traces of domestic fire found in these places date back to this time.

a period around 900,000 years ago or slightly later. Only some remains in Indonesia appear to be older (but here, again, either the specimens or the dates are unclear and/or unreliable).

All available information—which, unfortunately, is very scarce—suggests that the migrations to Europe and Asia were almost simultaneous. Evidence to the contrary may come to light in the future, but for now, fossil remains seem quite consistent.

The climates at that time suggest that Europe and Asia were going through an ice age. It is known that this lowered the level of the seas, creating natural land bridges in places that were previously submerged. Moreover, some areas in which there are deserts today were covered with vegetation at the time, allowing *erectus* to survive by hunting and gathering. The tropical forests in many areas were thinning out, making it easier to cross them.

With the onset of the interglacial period (which many place at around 500,000 to 700,000 years ago), the natural bridges receded and lush tropical forests abounded once again. Were *erectus* stranded in Indonesia? Is that why their remains seem more primitive today?

Although *erectus* had already started to settle down and transform, giving rise to local mini-evolutions, it is probable that there were later migratory flows in Asia and Europe (conceivably—why not?—in both directions). The remains of *erectus* in China are, in fact, somewhat different from those in Europe and northern Africa. The characteristics also vary with time in each of these regions.

Actually, the dates of the various prehistoric ice ages, especially the earlier ones, have not been precisely established. Interpretations vary according to the individual scholar, but new techniques have now come to the aid of classic "Alpine chronology" (that is, the study of the expansion and the contraction of the Alpine glaciers). Drilling in the Atlantic Ocean and the study of the oxygen isotopes contained in ancient strata or research being carried out in other areas, such as Poland, may lead to further insights.

While some scholars tend to set the exact dates of certain glaciations forward or backward, modify the intervals between them, or disagree about the number of phases, it seems reasonable, however, to assume that there is a link between climate and the migrations of *erectus*. Furthermore, there is a striking coincidence of certain dates: the use of fire in the northern regions inhabited by *erectus* appeared 400,000 years ago; at Zhoukoudian, 460,000 years ago; at Vertesszöllös (in Hungary), 450,000 years ago; at Arago (France), 400,000 years ago; and at Terra Amata (France), 380,000 years ago. A look at the climate at the time shows that this was the height of another ice age (the Mindel Glacial Stage, 300,000 to 500,000 years ago).

It may be hypothesized, therefore, that the daily use of fire originated outside of Africa, at the hands of hominids "pre-adapted" to its mastery and strongly motivated to improve the technique.

In fact, fire was not an "option" in the colder regions (as it might have been in Africa); it was essential. It was crucial for the populations in the higher latitudes during a glaciation to learn how to dominate fire. (The use of furs as protection against the cold probably also dates back to this period.)

We will return to the use of fire in chapter 8, in which we will consider the difference between knowing how to *use* fire (an ability which may be very ancient) and knowing how to *produce it when required*, a different matter altogether.

In the meantime, let's take a closer look at the travels of *erectus*, especially in an area that is of great interest to us: Europe and Italy.

The Most Ancient Europeans

The exact date of the arrival of *erectus* in Europe has not yet been established: it may have been around 900,000 years ago. Signs of their passage, some more reliable than others, have been found here and there, as will be seen later. Although investigations are progressing slowly, some results have been obtained.

The most ancient fossil, found in Heidelberg, Germany, is a mandible that dates back 650,000 years. It is an astounding jawbone, with rami that are twice as big as ours. Those *erectus* must have been incredible chewers; a bite like that would definitely have left a deep mark. They had practically no chin. Their teeth, however, were surprisingly similar to ours (even though the roots were longer and often fused).

We know nothing about these *erectus*, other than that they probably used tools that were much more rudimentary than those used by the Chinese *erectus* in Zhoukoudian (some tools from that time have been found in various sites).

The remains of the second most ancient European *erectus* were found near Budapest, at Vertesszöllös: they consist of an occipital bone and a few infant's teeth. From a hypothetical reconstruction of the skull, it would seem that the volume of the brain of these "Hungarians" was highly developed: over 1100 cc. In fact, they knew how to use fire;

remains have been found of hearths and tools (the time was, coincidentally, approximately 400,000 years ago).

These two places in Germany and Hungary are the most ancient sites in which fossil fragments have been found in Europe. But there are places in which indirect evidence of the earlier passage of *erectus* is present.

The most ancient (and reliable) of these seem to be the Vallonnet cave on the southern coast of France and Monte Poggiolo near Bologna, Italy. Stone and bone tools dating back 900,000 to 950,000 years were found in a small "room" fifteen feet from the entrance to the cave in France. At Monte Poggiolo, on the other hand, only stone tools, probably chipped by hunters from pebbles picked up at the beach, were found.

There is, however, some mystery about the tools made by *erectus* outside of Africa. The paleontologists call it "cultural phase displacement." It seems as though the *erectus* who migrated lagged behind their African counterparts in the technology of chipping stones by as much as hundreds of thousands of years.

In fact, in Europe the most ancient artifacts (around 900,000 years old) are not bifacial. In Africa, bifaces were in use at least 1.5 million years ago (Melka Kunturé, Ethiopia) or earlier (Olduvai). Why were the European *erectus* not aware of a technique that was so widespread in eastern Africa? Did they leave before it was invented? Or was it "lost" through the course of millennia because of limited interaction? Or have ancient bifaces simply not yet been found in Europe?

Whatever the reason, the fact remains that the earliest bifaces found in Europe date back 600,000 years (perhaps imported by a new migratory wave). This is another characteristic that must be clarified about *erectus*, and if it is, current theories could be modified.

But let's return to the cave in Vallonnet. Deer antlers were found lining the walls of the cave, but there were no signs of fire. Fossil pollens and animal remains indicate several changes in the environment: at the time of the presumed presence of human beings, the environment must have been a cold steppe with pine trees, populated by animals typical of such a climate, like wolves and bears. But in the strata dating back 1.3 million and 700,000 years, the remains of hyenas, rhinoceroses, elephants, and macaques have been found, indicating a succession of different climates.

32. The main European sites in which traces of *erectus* have been found.

The Côte d'Azur evidently already had a mild climate at that time; other *erectus* settled in that area in subsquent periods. Two other European sites are located in this region: Terra Amata (around 380,000 years old) and Lazaret (about 130,000 years old), also a cave.

Caves offered not only natural shelters but an environment conducive to the preservation of fossil remains, as they are less exposed to the transformations produced by human beings and nature (erosion, rain, floods, agriculture, etc.).

In other words, it is not certain that human beings always lived in caves. On the contrary, it is quite likely that they used other kinds of shelter. But the traces of their passage have been preserved better in caves than elsewhere. That may be the reason for the frequent and erroneous reference to our ancestors as "cave men." Caves were only one and definitely not the most frequently chosen places to camp during their nomadic wanderings (in fact, there was also the danger of finding unfriendly co-habitants, such as bears, hyenas, and sometimes lions).

We will return to Terra Amata and Lazaret in the following chapters. The abundant remains in both sites (huts, pallets, poles, hearths, tools, and traces of food) provide an extremely important source of information—almost an inventory of their everyday objects—on the lifestyle of *erectus*.

Some Prehistoric Residences

At this point, it may be a good idea to look at the "living sites" of *erectus* discovered to date and the remains attesting to their presence in Europe. One such site is the beautiful cave in Arago, in the Pyrenees, dating back 450,000 years, in which over fifty human fossil fragments have been found (mandibles, teeth, femurs, and, above all, an incomplete but very interesting skull: it includes the forehead and is, therefore, the most ancient European "face").

Thirty feet inside the cave is a small sandy depression where the small tribes probably camped.

The entrance of the cave, which is nestled like a castle into the mountainside, overlooks the entire valley once roamed by bison, musk oxen, and chamois, providing an exceptional vantage point for hunters. Some scholars believe that snares and nets were already used to hunt smaller animals like hares. A gorge leads up to a plateau where herds of deer once grazed.

Another cave once inhabited by *erectus* (dating back 200,000 to 500,000 years) has been found at Petralona, Greece, in an area full of underground tunnels and grottos. Huge stalagmites create a science-fiction atmosphere. In one of these huge grottos, called the "Mausoleum," an almost perfectly intact skull, protected by a fine layer of stalagmites, was found. The brain volume, 1200 cc, would suggest a rather evolved individual, even if the appearance, due to the brow and the receding forehead, is still very primitive (at least from our point of view).

Two more "Europeans" have been found, both dating back 250,000 to 270,000 years: one is "German" and the other "English." The English remains were found at Swanscombe, about 30 miles from London. They are of a woman approximately twenty years old with a cranial volume of probably more than 1300 cc. The German find at Steinheim, near Stuttgart, is also of a woman, with a brain volume of 1100 cc and more modern features, especially the shape of her cheekbones. Judging from the size of the nasal opening, she must have had a very large nose. Moreover, she may have had a violent encounter (maybe with a male) shortly before death because traces of a terrible injury on the left side of her skull suggest a hard blow with a dull object. But the process of fossilization sometimes plays tricks and it could simply be a deformation due to sedimentation.

While referring to different areas and different times, this brief review of the *erectus* in Europe shows a remarkable continuity in the physical model. They seem to have remained essentially the same over a very long period of time, though their brain volume increased. This development led to volumes of 1300 cc, as demonstrated by the skull found at Swanscombe (which, together with the one found at Steinheim, are considered forms of primitive *sapiens*).

And what about Italy?

The Mystery of Isernia

There is one Italian site that is very important in the study of *erectus*. It is among the oldest in Europe (over 700,000 years) and also one of the most striking: Isernia La Pineta.

As has often been the case in the history of paleontology, this find was discovered by chance, during construction of a highway on the outskirts of Isernia in 1978. Not unlike other times, an amateur naturalist, Alberto Solinas, noticed animal fossils and artifacts in the piles of earth being moved by the bulldozers.

Informed of the discovery, Professor Carlo Peretto, paleoanthropologist at the University of Ferrara, and Professor Benedetto Sala, paleontologist at the same university, along with supervising professor, Bruno D'Agostino, immediately inspected the area and confirmed the existence of a large prehistoric site (later dated to 736,000 years ago).

Isernia remains a mystery. Questions raised by the finds made there are still unanswered.

The excavations carried out to date have revealed two distinct areas: the first is approximately 680 square feet and the second about 1,200 square feet (the latter, however, is made up of two layers, the lower being approximately 400 square feet). Thousands of manufactured tools, but above all an enormous quantity of bones of large animals—tusks, humeri, scapulas, mandibles, and skulls (of bison, rhinoceroses, elephants, and even some bears and hippopotami)—have been unearthed.

Why is there such an accumulation of bones in this spot? Nothing of the kind has ever been found in any other prehistoric site. The sight of all these bones (now on exhibit at the museum in Isernia, thanks to a meticulous and careful reconstruction) leaves one breathless. The visitors (and researchers) leaning against the railing encircling the exhibit look in silent wonder.

Let's attempt some hypothetical solutions to the mystery.

Could it have been a slaughterhouse? Are these the remains of animals that were killed and then eaten? A number of arguments contradict this hypothesis.

1) It is difficult to think that the people of Isernia killed and *dragged all these animals to their camp.* Why should they have undertaken such a tiresome job (transporting the entire animal—bones, tusks, and all—for many miles) when they could simply have cut away the meat in situ?

2) The site at Isernia may have been a natural trap in which certain large animals got bogged down, thus providing the *erectus* with a constant food supply.

Study of the sediments shows that the site was once a swamp. Thus, one could hypothesize that these animals ended up in the swamp either accidentally (or driven by hunters), providing a ready meal for the *erectus* camped nearby (rather like spiders who feed on the prey caught in their webs). This could explain the presence of large, heavy bones. But why are there no complete skeletons, that is, ribs, vertebrae, etc.? One of the surprising things about the bones found in Isernia is the absence of "anatomical connection." For example, there are skulls but no mandibles of bison, tusks but no skulls of elephants, and so on.

3) The small tribes of hunters must have been extremely voracious if the remains were of a slaughterhouse (basically the garbage heap).

33. The extraordinary stretch of bones—of bison, rhinoceroses, elephants, etc.—found at Isernia. Its meaning is still mysterious. One hypothesis is that over 700,000 years ago, *erectus* paved the area around their camps in the marshes to make them more habitable.

Although it could have been used for that purpose over a long period of time (and, in fact, the site was inhabited in different periods), the idea that anyone could have eaten so much meat (and left the bones on the ground) is quite incredible. Nothing like it has been seen any-where else, even in more recent times. Furthermore, if it was a gar-bage heap, it would probably not look the way it does, like a paved surface.

This brings us to the second hypothesis: that of a bone pavement. The idea that prevails among scholars is that the mass of bones is an artificial construction, designed and built by *erectus*: the oldest pavement in the history of mankind. Thus, all the pieces (or the majority of them) were gathered in the surroundings and brought to the site as construction material.

In other words, the bones could be of various origin: some could be the leftovers of meals (and in fact some were broken for the extraction of marrow); others could have been brought there deliberately. The fact that the bones are mixed with pieces of stone and travertine supports this hypothesis; it is as if these ancient "masons" used all the materials at their disposal to create a terrace. But for what reason? Why build a terrace in the middle of the wilderness?

One response is that the *erectus* may have been more intelligent than we think; they may have "corrected" nature by consolidating the banks of a river, for example. Since it was useful to live near the water (in fact, almost all urban settlements in the history of human beings have been on the edges of bodies of water), it may have been worthwhile to improve the site by keeping the banks from sinking away into the swamp.

The advantages of living close to the water in Isernia may have compensated for the drawbacks of this situation. And then, the pave-ment may well have been covered by wood and sticks, of which no traces remain.

Some people have hypothesized that the material was used to mark the residential areas of the site (for example, a great many artifacts are concentrated in a certain place surrounded by pieces of travertine). There are also presumable traces of fire, that is, patches of reddened clay and bone that seem to have been subjected to a strong source of heat. But they provide insufficient evidence in themselves; should supporting evidence be unearthed, these traces could be the first definitive proof of the use of fire in that period.

Research goes on and could offer some surprises. The area that has been explored so far is minimal. Surveys have revealed that the site covers an area of 24,000 to 36,000 square yards; little more than 240 square yards, that is, one one hundredth of the total site, have been examined so far.

Therefore, much remains to be done and understood. In Isernia (as elsewhere), it is extremely difficult to put oneself in the place of individuals who lived so long ago. Professor Carlo Peretto who, along with his assistant Gianni Giusberti, has "lived" for years with these invisible ancient inhabitants of Isernia, has this to say on the matter.

It seems that the farther back we go in time, the less we are able to understand the remains left by our ancestors. I wonder whether that diagnostic ability is related solely to the time that has passed or whether it is also due to a more general lack of comprehension of human groups with requirements, behaviors, and lifestyles different from ours.

We too often are moved by the desire to find a logical answer at all costs to what we discover and study. In particular, I am referring to the "inhabited areas," above all the more ancient ones, excavated in Africa and Europe. Perhaps, as so often happens, we will never be able to discover why the artifacts found in Isernia were accumulated in that way.

In any case, the more we know, the more we realize that these primitive beings knew how to do many things?

Yes. And they were very skillful, too. The dates for certain abilities, formerly thought to be much more recent, are constantly being set back. I do not rule out some important discoveries in the future that would completely overturn present hypotheses.

Thus, it is hoped that new keys will be found to unlock the mystery of Isernia, but also that traces of the main character of the story, that is, the *erectus* who left the artifacts, will be found. The probability of finding human remains seems to be good, since the site is well preserved and, above all, was frequented at various times. The hope of finding *erectus* should encourage not only the painstaking work of researchers, but also the allocation of funds to support this extraordinary exploration of the past.

Erectus on the Peninsula

The find in Isernia seems to support hypotheses made in other sites with similar characteristics in Africa (Olorgesailie, Melka Kunturé), in France (Soleilhac), and elsewhere. The presence of various kinds of solid material could attest to intentional construction activity by *erectus*.

In this respect, the cobbled area found in Venosa, Basilicata, Italy, is particularly important. Many studies have been carried out by Professor Marcello Piperno of Rome, archeological director of the "L. Pigorini" Prehistoric Ethnographic Museum and expert on the populations of ancient Italy.

The site in Venosa was much frequented by *erectus*, albeit much later (400,000 to 450,000 years ago). The thirteen different levels of settlement found attest to its favorable position for hunting and to the "tradition" (handed down from generation to generation) to return there.

In addition to implements and animal remains, these thirteen levels have revealed an amazing number of pebbles, mostly split, chipped, or flaked. On one level in particular, the number of pebbles and the way in which they are arranged suggest that they were put there by someone (a pavement similar to the cobblestones found in certain old streets and courtyards). Of course, no proof exists that human hands arranged these pebbles in this way, but the extent of the find (thousands of square yards), the regularity of the pebbles (four to five inches in diameter), and the apparent lack of any natural cause justifying that kind of deposit suggest that *erectus* was somehow involved in the construction of this paved surface.

A femur (a diaphysis) was also found in Venosa, which has now been dated to 360,000 years ago, confirming that the site is very old. This find has a story to tell. Professor Piperno explains:

Examination of this diaphysis has revealed signs of a thigh injury, probably a wound which became infected, causing the growth of new tissue around the bone. It's extraordinary to think that this diagnosis could be made hundreds of thousands of years later by Professor Gino Fornaciari, a pathologist at the University of Pisa.

Isernia and Venosa are only two of the sites in which traces of the passage of *erectus* have been found in Italy. There are at least forty others. Some of these are in Fontana Liri, Lazio, Visogliano near Trieste, Fontana Ranuccio near Anagni (Lazio), Grotta del Principe near Ventimiglia, and Castel di Guido near Rome.

The last site has been studied at length by Professor Francesco Mallegni of the University of Pisa, in collaboration with Professor Antonio Radmilli. Dating back 300,000 years, it is the source of a number of beautiful bone and stone bifaces—extremely rare, almost unique artifacts.

Much pioneering work was done by Aldo and Eugenia Segre. Finds indicate that *erectus* lived throughout the peninsula. The most important finds are in the central and southern areas, but that may be the result of the greater difficulties created by alluvial deposits and the more intense agricultural, industrial, and urban development which has transformed the terrain in the north.

The Approach

The most recent sites of *erectus* found in Europe and elsewhere point to the approach of more and more human forms: *Homo sapiens* (although not yet *sapiens sapiens*).

The story will be told in the next chapter. It also holds some more mysteries for paleontologists, such as the emergence and disappearance of the Neanderthal people in Europe and the birth (probably in Africa) of modern human beings.

It took three million years for the first forms of biped hominids living on the African savannah to evolve into the *erectus* found in Isernia and Castel di Guido: a very long time, full of transformations and events which modeled the brain and the hand. It also made it possible for very ape-like creatures still living in trees to develop their intelligence; to become more skillful in working stone, hunting animals, and building shelters; to domesticate fire; to organize social and group life; and to mature from almost animal-like communication to the use of articulate and refined speech, able to express abstract concepts and transmit culture.

We have tried to outline the early steps of this adventure chronologically, that is, to follow the finds that have revealed this evolution, and to try to reconstruct the appearance of these hominids from the information provided by artifacts, fragments, and fossils.

Now one of the central problems in this evolution will be dealt with: the development of the brain (and therefore language), which allowed for the birth of intelligence and communication. In fact, the secret of the transition from animal to human lies in the development of the capacity of the brain. The upright position or the use of tools would have been useless without the sudden growth of cortical neurons which allowed mental faculties to flourish.

We will turn, then, to the investigation of brain development and how it led to the development of human society. The scant evidence available, but also the concrete knowledge gained from biology on the evolution of living beings in nature, will help us to reconstruct this all-important moment in human history.

The chapters that follow will be a kind of cross section of human history. The vertical sequence of events will be abandoned to describe what led to certain parallel (but related) developments that made the appearance of increasingly human beings possible: the development of the brain and speech, naturally, but also the mastery of fire, the construction of shelters, the manufacture of tools, the capacity for hunting, and social interaction.

It seems important to look into this subject at this crucial point in human history, which led to the increasingly rapid development toward *Homo sapiens*.

Let's start with the brain and with speech.

7

The Brain and the Word

Initial Data

In order to understand the rapid growth of the brain and intelligence in the human species, we have to start with a fact that is often overlooked: the brain is an organ like all the others. Like all the other parts of the body, it is composed of cells.

It is not surprising, therefore, that it underwent changes during the course of human evolution (like the teeth, the bones, the muscles, the joints, etc.) These changes can be seen everywhere in nature: animals have diversified a great deal over millions of years—and the changes are very apparent at times.

Furthermore, the brains of mammals, and especially primates, were already well organized; the kinds of neurons were the same as they are today in the various species—only the proportions were different. Intelligence developed in step with the development of the external layer of the brain known as the cortex, and in particular, those areas of the cortex that regulate associative activity.

But what brought on the sudden acceleration in the development of the brain of hominids?

It should be pointed out that this was not the first time that the brain made a "giant leap forward." The first occurred in even more remote times, between 230 and 700 million years ago, with a rapid increase in the mass of the brain followed by a slowdown. The English anthropologist Bernard Campbell has calculated that without that acceleration, the development of human beings might have taken 20

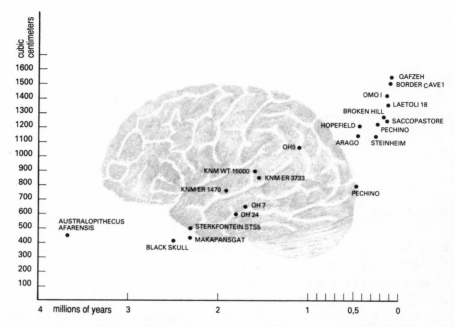

34. From Australopithecines to *sapiens sapiens*, finds show a gradual increase in brain volume. Plotted on this graph are only some of the skulls for which brain volume has been calculated. Of course, these remains belong to indviduals differing in height, sex, and species, and are, therefore, not directly comparable. An increasing trend is nonetheless evident.

(rather than four) billion years. This means that there would never have been human life on earth because the planet would have disappeared in the meantime, engulfed by the sun (this should happen in four to five billion years, if there are no complications).

Thus, the brain, like all the other organs, has had its spurts of development in the course of natural selection, not only in the line that led to human beings, but also in many others. In the last few million years, the tendency for brain mass to increase (resulting in ever more efficient brains) has taken place in many types of animals, from reptiles to birds and mammals.

It can be demonstrated that there has been an increase in the brain mass of ungulates (cows, horses, deer, etc.), carnivores (cats, dogs, foxes, etc.), and primates (all the apes and the prosimians) in the last 65 million years. This increase has been greater in diurnal than in nocturnal primates, in land-dwelling than in tree-dwelling primates, and in social than in solitary primates.

Human beings (diurnal, land-dwelling, and social primates) suddenly found themselves in a very particular situation: use of their brains started to be decisive for survival (cleverness, tools, strategies). Selective pressure for this characteristic began to be exerted.

Intelligence as an Instrument of Survival

One must be very careful here not to succumb to a popular misconception. Some people think that selection means a "fight for survival," in which the strongest or the most intelligent eliminate the others, perhaps even violently. The mechanism was probably quite different and much more subtle. The hominids that were more intelligent simply managed to live a little longer and, thus, have more children or protect them for a longer time. This meant a relative increase in the number of their descendants (generally with analogous features). And this in turn meant the success of certain genetic characteristics.

Things occurred in the same way in the animal world: those hereditary characteristics proved successful that were better adapted to the environment and thus allowed for increased reproduction. This is the way in which all the characteristics that we see today in nature arose (from fangs to camouflage, claws, speed, horns, and venom).

Using their brains as instruments for survival, at a certain point human beings passed a threshold beyond which intelligence became a factor in selection (that is to say, selective pressure). This led to the rapid development of mental faculties rather than horns, fangs, or camouflage.

It should not be forgotten that these mental faculties are enclosed in the two millimeters of the cortex, although an adequate support system had to be built up contemporaneously in other layers of the brain.

Attempts are now being made to measure these transformations that took place throughout the course of human evolution. We now weigh the brains of hominids (or rather, we calculate their cranial volumes on the basis of the remains of their skulls) and hypothesize that an increase in the volume of their brains corresponded to an increase in intelligence.

This is, of course, a rather risky supposition because it accounts only for certain factors and not for others, but for the moment nothing better is available.

The Volume and the Wiring

What counts in the brain is obviously not only the number of cells but the way in which they are organized. One example: the dolphin. If one considers the brain volume/body weight ratio, dolphins should be more intelligent than human beings because their brains are proportionately larger. But the development of the dolphin's cortex seems to have "specialized" in the processing of acoustic signals—an important characteristic, but not as important as those developed in certain centers of the human brain, such as speech and memory.

Fossil remains seem to show that some Neanderthals had very large brains—in certain cases even larger than ours—but does that mean that they were more intelligent? Probably not. "Less strategic" centers may have been more highly developed.

For example, some areas of the brain of modern human beings have receded in the course of evolution. One of these is the center regulating smell, the so-called rhinencephalon, which is usually highly developed in animals (especially in mammals, both predators and prey). As we all know, the rhinencephalon is very important in dogs, which live in a world of odors; a large part of their brain (in terms of both volume and weight) is dedicated to this sense.

Therefore, not only is the volume important, but also the areas in which this development has taken place, the organization of the cells, the so-called wiring of the nerve circuits, and the hierarchy of the various systems. Also important are the sophistication and ramifications of the connections, the biochemistry regulating their functioning, and the nervous system linking and coordinating them all. It is now known that many parts of the brain are redundant, that is, they carry out parallel jobs and can, if necessary, take over for others, so that the brain can function almost normally even at a reduced size.

A very famous example of this is a young woman who discovered at the age of twenty, after undergoing medical tests for some disorders, that the two hemispheres of her brain were completely independent.

A malformation at birth had prevented the linkages in the corpus callosum from forming and connecting the two hemispheres of her brain.

Accurate testing revealed that both hemispheres had developed and functioned totally independently of one another. Oversimplifying, each of the two 700 cc hemispheres functioned like a 1400 cc brain. In this regard, Professor D. Ploog of the Max Planck Institute emphasized that, "There are individuals with a 700 cc brain who can learn the basics of speech."

Another famous case involved two writers—Anatole France and Ivan Sergeyevich Turgenev. The former had a brain volume of only 1050 cc, the latter one of 1900 cc—two extremes probably due to exceptional conditions. (Harry Jerison of the University of California at Los Angeles, a scholar in the field, believes that both were pathological manifestations: the former of aging, the latter of hydrocephalus.)

How to Increase the Odds for a Good Poker Hand

The relation in evolution between brain volume and intelligence has been studied at length and has often been the source of heated controversy. Professor Ralph Holloway of Columbia University feels that volume in itself cannot be taken as a sufficient indicator of intelligence; what counts is the internal organization. Harry Jerison, on the other hand, agrees that organization is crucial, but feels that from an evolutionary point of view, the increase in brain volume can be considered an important indication of the complexity of the brain (and, therefore, in a very general way, of the evolution of the brain and of behavior).

A parallel might be drawn with a game of poker: it is obvious that the "wiring" of the cards, that is, the way in which they fit together (three of a kind, full house, flush, etc.) is all-important; but it is also obvious that if the game is played with two decks rather than one, or if every player receives ten cards rather than five, it's much easier to come up with three-of-a-kind, a full house, or a royal flush.

Impressions in the Skull

How was the brain of the first hominids structured? What kind of a cortex did the Australopithecines have? These questions are not easy to answer because we have no specimens of their brains: brains, like muscles or fat, do not fossilize.

What can be learned from the study of apes today is that the human cortex is proportionately 3.2 times as large. Basically, evolution has brought about this change in quantity and quality (it may be that at a certain point, as Jerison says, quantity triggers quality). Could the first steps in this direction already have taken place with the Australopithecines?

Professor Holloway feels that there is evidence to support the affirmation that the external layer of the brain was already being reorganized in the Australopithecines. In particular, the posterior parietal region, that is, the area in charge of integrating visual stimuli with other stimuli, was undergoing enlargement. This was to play an important role in the development of social behavior. In fact, Holloway believes that the development of a more complex brain started with the emergence of the upright position three or four million years ago. In other words, Lucy may have already shown the first symptoms of the quantitative and qualitative growth that began with *habilis* and blossomed with *erectus*.

But is there no direct or indirect fossil evidence of brains? Some feel that there may be: the impressions left by the brain on the inside of the skull. They not only indicate the general form of the lobes, but may also provide indirect information on other structures.

Phillip Tobias comments:

> The brain models the skull during its development. With each heart beat, it presses against the walls of the skull, leaving its "impression." The signs of the convolutions of the brain are more pronounced in young people and it is fortunate that most fossil remains belong to people who died at an early age, the best time to leave these impressions.

Ralph Holloway also feels that the study of these "fingerprints" of the brain can reveal some aspects of its organization. Furthermore, he is convinced that the basic structure of the human brain had already been established two million years ago by *Homo habilis*.

Tobias goes so far as to state that these similarities emerge quite distinctly with *habilis*:

As far as the frontal lobes—so important in human beings—are concerned, Australopithecines are still very ape-like, while the convolutions of the frontal lobe and the Broca's area [controlling speech] of *habilis* are already arranged like ours. The parietal lobe, which is essential for thought and hosts the area of Wernicke, is naturally very highly developed in human beings, but the inferior part was already well developed in *habilis*.

There seems to be (cautious) agreement among many researchers today that certain signs of the evolution of the brain can be detected from the shape of the skull and the impressions left within it. Some are, however, more skeptical, stressing that things are not that simple.

Professor Erik Trinkaus of the University of Albuquerque, New Mexico, says:

The form of the convolutions cannot be determined from the impressions left on the skull. The grey matter is rather thick and only the impressions of the blood vessels of the walls of the skull are evident. Those of the brain lie deeper. It is true that the development of the frontal lobes and things like that can be seen, but as far as speech is concerned, for example, only Broca's Cap is visible and that has little to do, neurologically, with the Broca's area which is involved in speech.

Speech: The Scriptwriter and the Puppeteer

A little geography of the brain may be in order here to explain what the areas of Broca and Wernicke are.

Broca's area is responsible for speech in the sense that it coordinates the nerves that move the muscles of the mouth, the tongue, and the pharynx. This center is something like a puppeteer who maneuvers the vocal apparatus like a marionette to make it produce sounds.

But puppeteers do not write the script, they only perform it. The dialogue originates in the area of Wernicke which is, in a certain sense, the real center of speech in that it organizes sentences and paragraphs, grammar and meaning. This is the origin of the orders which Broca's area encodes into nerve signals that are, in turn, translated into sounds and words.

On the one hand, therefore, Wernicke's area creates ideas, thoughts, structures, and language and, on the other, Broca's area encodes these ideas into motor signals that activate the vocal apparatus. Both areas are obviously necessary for speech.

We know that both exist in human beings, but to what extent were they already present in early hominids?

Since no brains exist from that period, and given the difficulty in deciphering the impressions left in skulls, another source of comparison is necessary. It may be useful to examine the brain of the chimpanzee to see whether there are traces of the area of Broca in such evolved animals.

Professor Jürgens of the Research Department on Primate Behavior of the Max Planck Institute in Munich says that the area of Broca exists in chimpanzees. It is very small, however, and does not seem to have speech functions; it seems to control other functions which have not yet been understood. Professor Jürgens feels that this Broca's area (which probably already existed in the ancient primates) underwent a functional change during evolution.

It would not be the first time that something like this occurred in evolution. As French biologist and Nobel Prize winner François Jacob put it, the evolutionary handyman often uses old pieces to create new things. Just as an old wheel makes an excellent fan or a sun shade can be fashioned from a table, so was an esophogeal hernia transformed into lungs. In reconnecting circuits in an increasingly elaborate network, Broca's area may have been reorganized in a new way, linking with speech mechanisms as a pre-adaptation.

The Animal Sounds of Human Beings

It is interesting that this evolution affected only certain functions. Our "talking machine" is, in fact, made up of various parts, some of which are ancient, others more recent. The primitive vocal systems, typical of animals, are still present in human beings and become evident under certain circumstances. For example, Jürgens explains that there is a tiny motor area behind the area of Broca, the destruction of which causes paralysis of the facial muscles in both human beings and animals. If destruction is bilateral, human beings become mute, apes do not. After

35. The centers of speech and phonation are quite different in human beings and in apes because their brains are different (and, therefore, also their intelligence). But human beings still have vestiges of archaic vocalization systems that can be called into operation in case of lesion.

a few weeks, however, human beings start to produce sounds again, using an ancient structure to vocalize in a pre-human manner, so to speak, that is, without modulations. In this way, they can express only emotions, as apes do.

Another important discovery has also been made. If certain lesions occur, these ancient sounds can no longer be produced voluntarily. The individual can only scream if subjected to pain but, like an animal, cannot scream voluntarily.

With other kinds of lesions, the power of speech may remain, but the capacity for intonation and emotional coloring is lost; the individual speaks like a machine.

Thus, speech seems to be the result of various parts that developed in different evolutionary stages. Some of these parts have remained essentially the same and are similar to those of animals, while others have developed more recently and have integrated with areas of the brain that were also undergoing transformation.

This is why it is incorrect to speak of a single center of speech in the brain. A number of systems and subsystems work together to produce speech. They include intelligence, emotion, the ability to encode and transform words into motor impulses, the mechanical adaptation of the larynx, and the vocal cords.

Just as the transition to the upright position involved the gradual transformation of a number of parts of the body (the knee, the hip, the spine, the attachment of the skull, etc.), the mastery of speech called for the transformation of various parts of the brain and related systems, such as the vocal apparatus.

In fact, a well-functioning brain is not enough for speech; it takes a system capable of articulating sounds—a loudspeaker, so to speak.

Chimpanzees, for example, lack suitable loudspeakers. Even if they were intelligent enough to learn a few words, they would never be able to pronounce them (or they would pronounce them in a different way). Human beings, on the other hand, have an efficient system for translating thoughts into words (it may be used to express insane thoughts, but they are nonetheless well articulated!).

Breathing and Drinking: The Evolution of the Larynx

An American researcher of the Mount Sinai School of Medicine, Jeffrey Laitman, has used what might be called paleolaryngology, that is, the study of the vocal structures of our ancestors, to carry out an interesting investigation into the mechanisms of this loudspeaker.

The investigation started out from an interesting fact: animals are able to drink and breathe at the same time, human beings are not. Try to breathe while drinking a glass of water and you will see for yourselves what happens; at best, the water will "go down the wrong way."

This is attributed to the particular structure of the larynx, which is placed much higher in the neck of animals. However, this characteristic limits the ability of animals to modulate the sounds generated by their vocal cords. In human beings, the larynx is low in the neck, creating a large "resonance chamber" near the pharynx allowing for modulation of sounds.

It is interesting to note that newborn babies are much like chimpanzees and other mammals in this respect. Their larynx is still high. A study carried out by Professor Laitman using cineradiography shows that babies essentially breathe, drink, and vocalize in the same way that apes do, and that the structure of this part of their respiratory system is very similar to that of the ancient primates. After approximately a year and a half, however, a radical change takes place: the babies' larynx starts to sink, changing their way of breathing, drinking, and vocalizing.

This change can cause serious problems (a crumb blocking the entrance to the larynx may result in suffocation), but in return it offers the enormous evolutionary advantage of being able to modulate sound and, therefore, articulate speech.

Professor Laitman has discovered that the shape of the base of the skull provides important information about the position of the larynx: while that of mammals is basically flat, that of human beings (but only after two years of age) is arched.

Using this criterion, he has studied the bases of the skulls of the various fossils of hominids available in African and European collections. He has discovered that the base of the skull of the Australopithecines was flat, similar to that of chimpanzees. Therefore, their vocal apparatus must have been similar to that of chimpanzees. They were able to breathe and drink at the same time, but were probably unable

36. Apes (and newborn babies) can drink and breathe at the same time; adult human beings cannot. The lower position of their larynx prohibits this, but in return allows for the articulation of sounds. Examination of skulls has shown that Australophithecines had a vocal apparatus similar to that of chimpanzees and, therefore, could probably not articulate sounds.

to modulate sound. Laitman believes that the Australopithecines had some sort of communication system, perhaps a little more advanced than that of apes today, but that they were unable to speak as we do.

When did the first human-like structure appear? Laitman's study indicates that it is with *erectus* that the base of the skull started to arch, and thus, that it is with *erectus* that the larynx started to descend, making modulation possible. The *erectus* could probably not articulate sounds well either, but their speech was probably already somewhat like ours.

Only much more recently—30,000 to 40,000 years ago—with the emergence of *Homo sapiens sapiens*, who were definitely able to produce speech similar to ours, did the base of the skull of our ancestors become arched like that of human beings. Recent research on the Neanderthals, carried out on the remains of a hyoid bone and its related brain areas, suggests that the Neanderthals were already able to articulate sounds well (this will be discussed further in chapter 8).

The Assembly of Phonemes

Of course, the position of the larynx is not the only factor determining the ability to articulate sounds; there are many others. Henri de Lumley, for example, points to the fact that the Australopithecines had a flat and rather shallow palate, which did not allow the tongue adequate movement. But *habilis* already had a deeper palate, suitable for articulation.

Another extremely important factor is that human beings have a much more sophisticated muscle system with a rich network of nerve fibers by which the tension of the vocal cords (and other parts connected to speech) can be regulated more precisely. As mentioned earlier, in fact, the vocal apparatus is only a puppet theater in which the strings, pulled by the brain, act out scripts prepared in the brain.

This is why no animals have ever been able to develop speech, regardless of whether they have a low larynx or a high palate: apes can learn a few words in sign language (but they do not seem able to compose sentences); dolphins can even manage some grammar and simple word combinations (but cannot express themselves). As Bertrand Russell once put it, "Although dogs and cats (or chimps and dolphins) are eloquent in their own way, they will never be able to tell us that their parents were poor but honest."

Human beings can. Just as it takes only a few notes to be able to compose all the music in the world, or ten numbers and a few symbols to work out the most complicated mathematics, with only about forty phonemes (basic sounds), human beings can express an infinite number of different ideas.

When did this ability to combine phonemes appear?

"I believe that speech started with *habilis*," says Phillip Tobias. "I have been saying this for years, but many people are now starting to believe it."

What was their speech like?

It may have been based on sounds similar to those used by the great apes today, but in different sequences. The great apes have different sounds for different situations and they just repeat them: "eh, eh, eh, eh, tsk, tsk, tsk," etc. I think that *habilis* used simple sequences of these primitive sounds to form short words like "abeh, abeh, abeh." If this ability is combined with the ability to make other sounds, then we have the birth of oral communication. But language is, of course, a product of thought. Therefore, intelligent communication is linked to the growth of the cerebral areas that are associated with the development of intelligence.

The Human Infant: Evolution Revisited

The transition from simple sounds to articulate speech actually took quite a while (although only a short time in evolutionary terms). And yet each of us can witness the same transition firsthand in just over a year: from the time a child is born to the time it learns to speak.

Some people feel that in those few years of maturation of the brain, the newborn child goes through a kind of recapitulation of the evolution of speech. It is well known, in fact, that during the nine months of pregnancy (gestation), the fetus seems to go through the various stages of evolution (from the first forms that lived without oxygen, as the fertilized ovum does; to fish, with a trace of gills; to reptiles, mammals, and human beings). Some people feel that the same thing occurs with the development of speech.

Of course, the idea of recapitulation is very controversial and must be taken "with a grain of salt." It is, however, striking that the larynx of the newborn descends in a year or two from its location in chimpanzees to its location in human beings. Furthermore, the development of speech in babies goes through the same stages that may have accompanied the evolution of speech in hominids, that is, only sounds of emotion at first, significant sounds later, then repetition of single phonemes followed by combined phonemes and, finally, complete speech.

Does that mean that babies have to be observed in order to understand the prehistory of human beings? Of course not. But we may be able to draw some information from this mini-evolution when more is known about the development of the brain during infancy.

A Growing Fetus

Naturally, a developed brain and an adequate vocal apparatus are not enough for speech and thought: a suitable cultural context is required.

Feral or "wolf" children, that is, children left to themselves during early infancy, can barely speak or think (this happened in Boston, where twins—the children of an alcoholic woman—were left untended for two years to grow up in a "savage" state). Parental care, education, a sense of belonging to a group and a culture are needed for individuals to develop their intellectual and linguistic abilities.

But only with the first hominids were the prerequisites for that development established, thanks to a special advantage they enjoyed: a long infancy.

Human infants are, in fact, very immature at birth; the human brain continues to develop and mature for many years. At birth, the human brain is only one fourth of its full adult size. Not only that, many neuronal systems do not yet exist. The curious thing is that (unlike the primates) the brain of the newborn child continues to develop at the fetal rate for the first six months, as if it were still in the mother's womb.

In that regard, the biologist Steven Jay Gould has calculated that if the various phases of human development (maturity, independence, and sexuality) are compared to those of other primates, pregnancy is relatively short. It should actually be six months longer, that is, fifteen months in all.

This is why some people feel that the human baby is still a fetus even after it has left the womb. Therefore, a gestation period of only nine months may be an evolutionary trick to allow the head of the baby to pass through the mother's pelvis.

This long external maturation exposes the brain of the "fetus-baby" —which is extremely sensitive because still being formed—to the influence of the environment, enriching it at a time when its capacity for adaptation is at its peak. But the long maturation also provides another advantage: it allows for a long period of parental education (and therefore cultural transmission).

Chimpanzees and gorillas also have long infancies and enjoy parental care for a long time. But here the mother and the child cannot converse as they lack the brain structures and mechanisms for speech. But when two (or more) beings are able to communicate experience, stories, and learning (even if only in a simple manner) and are together all day long for years, a culture is gradually formed. This is a decisive factor in improving survival and, therefore, the success of reproduction.

Speech is used everywhere: not only in the transfer of information, but also in the organization of work; in the setting of rules; and in the expression of desires, feelings, and plans. It is a formidable factor of cohesion in a group and can really unite people. While the brain is a set of neurons, culture is a set of brains, making possible a new superstructure not envisaged by genetic codes—a superstructure no longer dependent on neural connections alone, but also on cultural ties.

This structure can, of course, disintegrate, as it is not transmitted by chromosomes, as are hereditary traits. It is only "software." But it is so effective that it can reproduce itself through symbolic transmission via oral (and also written) language.

Thus, the development of the brain and of speech have been intertwined throughout human history, creating the selective pressure which probably gave our ancestors an intelligence and linguistic ability similar to ours about 30,000 or 40,000 years ago.

The Hand and the Tool

The development of the brain also led to another characteristic typical of human evolution: the ability to use the hand in a more sophisticated way. The hand has, in fact, become an extension of the brain, an external tentacle capable of translating thought into movement. It is operated by remote control from the brain by impulses delivering precise instructions. Just as a robotic arm depends on the computer

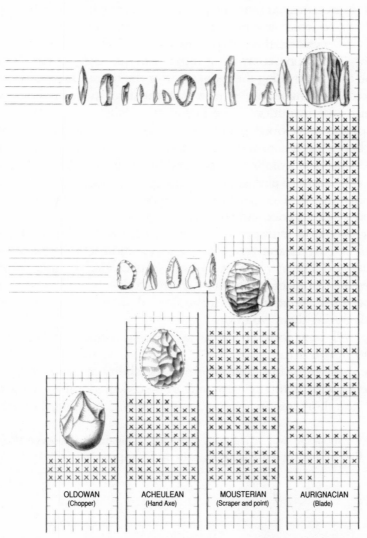

37. *Intelligence and tools.* The crosses indicate the number of blows (during working) needed to produce a tool: from just a few blows for a simple chopper of two million years ago, to almost 250 blows to produce a blade 20,000 years ago. The "mental" technique is also different: in the beginning, the chips were discarded; in the end, the flakes were the tools. Numerous other tools could be made from the same core (with further blows) and the "cutting edge" obtained from one stone was increased a hundredfold.

to which it is connected (and the program directing it), the hand depends on the brain to which it is linked (and the culture directing it).

It is no wonder, then, that a vast area of the motor cortex (that is, the brain neurons controlling movement) is reserved for the hand. Only the area controlling the movement of the mouth is equally vast.

Speech and hand movements are the two main activities which distinguish our evolution. Speech and the ability to manufacture objects probably developed parallel to the growth of the brain, and the two probably had a constant and reciprocal influence on each other.

Studying the evolution of the ability of the hand to manufacture objects and tools is almost like studying the evolution of the brain, since tools are, in a certain sense, a byproduct of mental activity—a reflection of intelligence and culture.

Besides the faint impressions left on the inside of their skulls, the hominids left far clearer traces of their evolution: chipped stone tools. Appendix 4, "The Advent of Tools," is dedicated to this subject. It describes the results of some very detailed studies carried out on various kinds of tools, depicting the technological (and mental) itinerary of hominids during their evolution.

We will now look at *erectus* at a magic moment in history: the discovery of fire.

The mastery of fire (and above all the ability to produce it) meant the invention of a totally new and unprecedented tool for *erectus*. It opened new horizons in defense, nourishment, hunting, heating, and travel.

Let's call upon our imaginary eyewitness again to observe the lighting of a fire in a camp of *erectus* in central Europe.

8

The Discovery of Fire

A Spark in the Hut

The sun is low over the horizon as the last touches are put onto the hut. Its oval shape and shadow make it look much longer than it actually is. Several people are moving around it. Their shadows stretch out long and thin on the ground. The members of the group have become accustomed to this barren landscape with its sparse shrubs and thickets, so different from the one they left behind up north only a few days ago. The elders say that the vegetation has thinned from past years. For the first time this season, they can see their breath before the sun goes down. That or the fact that the cold bites down to the bone may be why a young man is attentively watching his father crouch to make a shallow depression in the ground in the middle of the hut.

The ground is still warm.

The hollow is no deeper than the palm of a hand. The boy is seated opposite the entrance and can see the sun—a ball of fire—balancing on the earth's rim. It is blinding and he squints to look at it from under his heavy, pronounced brow. He tries to figure out what his father is doing silhouetted against the light. The boy can see his father's hand reach out and rummage in a bag, but what he pulls out is indistinguishable.

Are they some sticks tied together, or a long stone? All he can make out are his father's elbows occasionally protruding.

His movements are slow; he is obviously still in the preparatory stages. His hand reaches out to the bag once again and this time extracts

151

what seems to be a bunch of dry grass and moss. Now his movements are more regular; his elbows emerge at regular intervals.

Puffs of vapor form as the man's warm breath meets the cold air. The cold intensifies as the sun starts to slip below the horizon. The boy turns around. Almost all the other members of the group are seated around him, attentively watching his father's regular movements at the center of the hut.

Just as the sun goes down and almost as though to bring it back into the hut, a feeble light glows on the ground in front of the man.

He immediately bends down to blow gently at it. With each breath, the light becomes more intense. The father's face seems to materialize each time. Suddenly a tiny flame illuminates his smile. His strong hands place some dry grass on the tiny flame which responds with a friendly crackle. A thin ribbon of smoke rises as the flame is nursed with more dry grass and some kindling. A small torch is formed and placed under a bundle of twigs and branches lying in the depression. The light and the crackling increase. The flames spring higher through the wood, engulfing the larger pieces. The boy can now feel the warmth on his face and body. As if by magic, the group becomes animated and breaks out in excited chatter.

Someone laughs when a small branch pops, sending sparks shooting to all sides. A cold wind is blowing outside, dispersing the fine column of smoke rising from the hut. Far away, in the mountains, the temperature is already below freezing. But no matter, they'll all be warm tonight.

38. Hominids were probably familiar with fire from the earliest times: burning sticks were pulled from the smoldering ashes left by natural fires and used to light other wood. But producing fire when required was another matter altogether.

Lightning and Volcanoes

The preceding scene has been described with a deliberate lack of detail because of the scarce information available about the first fires. The study of the domestication of fire is especially difficult because only a few meager sources of information exist.

Investigators have some reliable data to go on (some fire sites and their approximate dates). They know what techniques *erectus* could have used to light a fire and know that some food was cooked. They have found depressions used as braziers for heating. But that's about all. The rest is yet to be discovered.

Let's see whether we can use logic and the little that is known to reconstruct other parts of the story. Lighting a fire is only the final act of a long process, which started much earlier.

We could begin by saying that neither *erectus* nor their predecessors invented fire; nature did.

Fire sites, that is, the remains of natural fires (lightning, spontaneous combustion due to the overheating of compact and humid vegetation and, of course, volcanoes, the lapilli of which ignited brush and woods), existed on earth hundreds of millions of years before the advent of human beings.

Millions of years ago, the first hominids appeared in a very volcanic area, as the footprints at Laetoli testify. Thus, from the earliest times, our ancestors were familiar with the strange and fearsome phenomenon of fire. Many probably fell victim to it, just as many animals were probably trapped in it.

The mastery of fire was, therefore, a gradual process. The sequence may have been as follows:

- taking lit branches from natural fires

- putting these branches into a hearth and keeping the fire going with other wood

- carrying lit branches when on the move to be able to rekindle a fire

- directly producing fire with specific instruments.

It is difficult to say how much time was needed to pass from one stage to another. However, evidence found during excavations makes some conjecture possible.

Let's start with the first point: using lit branches taken from natural fires. The first question is why? Why should a hominid extract an incandescent branch from a fire? Or stick a dry branch into a dying fire to light it?

It should be emphasized that a fire (which might last days or even weeks, and spread over vast areas) was a terrifying, but also an alluring event. It drove large numbers of animals into the open and therefore represented an excellent opportunity for hunting (many injured or suffocated animals could be captured more easily). Furthermore, the carcasses of burned or asphyxiated animals could be found where the fire had already burned out.

This suggests one plausible scenario. A group of hominids (perhaps the ancient *erectus* in Africa) approaches an area in which a fire has just gone out. While some people search for carcasses, others (perhaps some younger hominids) play with the last smoldering branches, poking at some coals with a stick. They have played the game before and have learned to be careful; after a while, the point of the stick catches fire and can be whirled around. At some point, they realize that by carrying a lit stick with them, they can light others later to make another fire.

Children today can be imagined playing a game like this. Could it have taken place in the past? Is it possible that fire was first domesticated unintentionally, without actually planning to use it later? The hypothesis seems reasonable.

Subsequent Steps

Once this initial ability to manipulate fire was achieved, it could be used for practical purposes. A multitude of these exist, but it took millennia of evolution to discover them.

As the hominids were unable to produce fire, they had to preserve it. That is why the first hearths were probably endowed with great importance: the fire had to be kept burning night and day and was never allowed to go out. (This rite of keeping the fire burning remained a tradition, although with different significance, in many cultures—from the vestals to the Olympic flame.)

But how could fire be transported when continuously on the move? (Our ancestors were essentially nomads.) The solution is simple enough. One example is provided by certain Pygmie groups (also nomads), who live in the forest: one member of the group carries a large lit stick.

But of course, the real revolution in fire came with the discovery of its production. And this discovery immediately solved all the other problems.

We now know that fire can be produced in a number of ways, but we do not know which was invented first. The striking of iron pyrite with a flintstone seems to be rather recent (certain evidence, found in a cave in Belgium, dates back only 10,000 to 15,000 years). The other two methods are the "classic" rubbing of sticks, and sparks flying from the chipping of flintstones onto hairs or dry grass. (There is some controversy about the latter because the heat from sparks in not enough to generate a fire.)

Whatever the first "matches," at a certain point in their evolution, our ancestors were able to reproduce what only lightning, volcanoes, or biochemistry had occasionally been able to produce before. The fire sites found in the camps of *erectus* provide clear proof of this.

But what do fossil remains tell us about the "chronology" of fire?

The First Traces of Fire

One fact stands out in a study of the African sites in which traces of fire have been found. Traces of combustion appear in two distinct eras: the first around 1.2–1.5 million years ago (Koobi Fora and Chesowanja in Kenya and Gadeb in Ethiopia) and the second 800,000 years later, between 350,000 and 400,000 years ago (Bodo in Ethiopia, Kalambo Falls in Zambia, and Cave of Hearts in South Africa).

A gap of over 800,000 years is odd, to say the least, since many *erectus* sites of that period have been found. Is it possible that fire was mastered over one million years ago, forgotten for more than 800,000 years, and only rediscovered about 400,000 years ago? Or are the finds that confirm the continuity of its use merely missing?

Examination of the most ancient traces of fire in Africa (those dating back to over a million years ago) raises some doubts. The fire sites are not, as they are in Zhoukoudian, very distinct and the finds could well be explained otherwise. They could simply be remains of natural fires which blackened the earth and the surrounding rocks.

Alan Walker comments on one of these sites:

I worked at Chesowanja; as a matter of fact, I was among the first to go there as soon as the remains of an *Australopithecus boisei* were found there. The first day that we were exploring the area, we found some strange objects—pieces of burned clay—here and there. The inhabitants of the area explained that these were the remains of a huge fire that had razed the entire area before the Second World War. Well, the site is right in the middle of that area. The possibility of contamination is too high. I don't believe it.

The same doubts exist with regard to the other sites and many researchers are quite skeptical. "No traces of fire have been found dating back more than 400,000 years," says Henri de Lumley. "After that it was used constantly."

A recent discovery in a cave in Swartkrans, South Africa, seems to shed more light on the matter. Of almost 60,000 fragments of bone found in one layer of a cave, 270 show traces of charring. The layer dates back 1–1.5 million years. Is this proof that the use of fire dates from that time? It could be, but there is much room for doubt.

Analysis of the bones shows that the burns were produced at different temperatures (some at 200°, others at 300–400°, still others at over 500°). Furthermore, previous studies have shown that these caves were used not only by hominids, but also by animals which dragged their prey (or carcasses found on the savannah) there so that they could eat them later. Could these burned bones not belong to animals that died in fires on the savannah? It's likely. Furthermore, the very fact that only 0.5 percent of the bones show signs of combustion would seem to indicate that fire was not used regularly (there were no traces of hearths).

Whatever the facts, it seems reasonable to assume that the use of fire was achieved gradually . . . and somewhat accidentally.

It may have been used occasionally at that time following natural fires (perhaps by keeping them lit for some time), but evidence of real domestication and use is lacking. This may also be due to the fact that, unlike stone tools, traces of fire sites—especially out in the open savannah—have not been preserved.

As it is, there are no traces of fire or even charred bones (with the exception of those found at Swartkrans) dating back more than 400,000 years.

Fire Produced by *erectus*

It's curious that the most ancient and reliable evidence of fire has so far been found outside of Africa—in China, Hungary, and France.

This does not necessarily mean that the use of fire was invented by the *erectus* who migrated. It merely means that fire was indispensable in those cold climates and that *erectus* (or at least some of them) probably already knew how to handle fire when they left Africa and applied this knowledge intelligently when it was required. And this was, of course, required when the ice ages set in. *Erectus* were, thus, pre-adapted to the use of fire: necessity drove them to improve on their techniques until they discovered how to produce it (a discovery that does not seem to be too remote).

Those environments, so different from the savannah, probably accelerated the regular use of fire, gradually transforming it into an indispensable daily tool. But was it its warmth that made fire such a success?

The use of fire actually provided a number of advantages. Some spring to mind immediately, while others are less evident and were probably only discovered in the course of time (and perhaps independently by various groups of *erectus*).

Although these hominids were used to living among wild animals, a fire in the camp kept predators at bay (perhaps involuntarily at first), especially at night (when they are most active). This fear of fire could also have been used to drive hyenas away from carcasses or bees away from a hive. As soon as fire became a constant in the camp, its contributions in the culinary field were discovered. This may have occurred quite accidentally when someone tasted the roast meat on a bone thrown into the fire, or intentionally, in an attempt to imitate what happens in nature when animals are caught in fires. Cooking kills parasites and eliminates toxins and was found to make some meat more digestible. It also makes meat more tender and less rubbery, providing children and the elderly with a nutritious and easily chewable source of food, thereby increasing their chances of survival.

It may also have contributed to some anatomical changes. "Eating cooked rather than raw meat no longer required as much muscle work," claims de Lumley, "and this may have led to a lighter facial and cranial structure."

But above all—and this can be easily surmised—fire had a fundamental effect on the development of human culture.

Beside the Hearth

For the first time, fire offered the hominids a few hours more light in the evening. They could use that time to prepare their tools, improving

39. The domestic use of fire greatly improved the way of life, nutrition, defense, and the migratory capabilities of *erectus*. It probably also contributed to the development of language and social behavior.

known techniques and discovering others. Bones and antlers, for example, could be hardened in the flame and used as hammers (this seems to have been the case in Zhoukoudian). The same could be done with green wood; instead of being burned, it was hardened and used to make more efficient spears.

At a certain point our ancestors discovered that flintstones could be cut and chipped better when heated, producing flatter and thinner tools.

But above all, the time spent around the hearth had become a magic moment for communication and socialization.

As already mentioned, *erectus* probably had a brain structured for relatively complex speech. The evening hearth was the most natural place for plans to be made for future hunting expeditions, for stories to be recounted about past hunts, and for roles to be defined. Not only that, it was the only time after a long day of hunting and gathering in which all the members of the group came together. It was the best time to deal with problems, to settle controversies, and to exchange experiences.

Speech was certainly still very simple and primitive; the vocabulary was limited. It took hundreds of thousands of years to refine it, through the long and reciprocal influence of natural selection and culture. But fire probably played an all-important role in this evolution. It enhanced one of the most typical characteristics of human development: sociability and, therefore, cooperation, an indispensable instrument for organized group life.

The *erectus* who lived in Europe 400,000 years ago used fire regularly in their camps. Traces of it have been found, for example, at Terra Amata, near Nice, in a site close to the sea, together with the remains of some huts. (Some of the huts were probably built later by the same groups of hunters who returned periodically to that locality, possibly because it was a good seasonal hunting ground.)

Thanks to a number of characteristics that later led to the gradual emergence of increasingly organized social groups (brain development, speech, tools, huts, and the use of fire), *erectus* started to hunt in a way that was to witness a strong future development.

As their hunting and, therefore, their predatory abilities improved, these hominids included more and more meat in their daily diets. They became predators of animals larger than themselves, much larger—even as large as elephants.

The next chapter starts with the report (based on available information) of an elephant hunt near Nice. As they say in the movies, the characters in this story are fictitious. But in this case, all references to situations are not merely coincidental, but are based on events that actually took place.

Let's go back to Terra Amata on the Côte d'Azur, to a dawn 380,000 years ago.

9

Hunting and Food

A Day at Terra Amata

The sun has not yet come up over the limestone cliffs girding the bay. Waves gently wash up onto the beach.

The contours of a thatched hut on the beach slowly become evident in the monotonous pale grey light. The red glow of the dying embers can be seen inside the hut. Some sleeping bodies, half-wrapped in furs, still lie motionless around the hearth.

Awakened by the cold, a stocky human figure hesitantly leaves the hut. The hominid looks up at the clear sky; it's going to be another beautiful spring day. Then, looking toward the sea, his attention is drawn by something on the beach: the antlers of a deer killed last year and found while rebuilding the hut in exactly the same place this year.

Other figures are slowly leaving the hut, stretching in the chilly air. A few children burst outside, chasing one another. Some adults set off to look for wood to liven up the fire, others sit down to eat some berries picked the day before. A mother lovingly caresses her child as she nurses it.

In the midst of the lush Mediterranean vegetation typical of this part of the coast, a spring of fresh, clear water can be heard gurgling not far away. Almost everyone has gone to drink from it. The sun will soon be up; this is the best time to set out on the hunt.

A few more berries and some dried meat complete preparations. Long wooden poles with tips that have been hardened in the fire are checked and a few last-minute touches are given the stone tools made the day before.

161

40. ". . . Awakened by the cold, a stocky human figure hesitantly leaves the hut. The hominid looks up at the clear sky; it's going to be another beautiful spring day. . . ."

The floor of the hut is strewn with chipped stone. The remains of a fish similar to a gilthead and the empty shells of mussels and oysters lie in the sandy hearth in the middle of the hut. The oval-shaped hut is solid, made of wooden poles covered with leafy branches. Although much larger, the hut is shaped like a pup tent. It does not keep out the wind entirely, but it does provide shelter from a climate which is much colder than it is today.

The hunters gather. Using gestures and words, they go over the strategy worked out the night before around the fire. The maneuvers are actually quite simple: surrounding the prey, cutting it off, and other simple but effective tricks.

After taking leave, the armed men silently move behind the more expert elders. Two young men on their first hunt are also in the group.

Beyond some Mediterranean scrub, the group turns onto an invisible trail leading through lecci and Aleppo pines and known only to those who return to this area year after year. The light filtering through the leaves of the oak and chestnut trees becomes brighter as dawn nears. The hunters silently quicken their step. An owl hoots in the distance.

It is surprising to see trees like birches, beeches, and alders at sea level. But this climate (halfway between a cold era and a warm one) produced vegetation along the coastline which would only be found in the uplands today.

The group laboriously climbs a rocky slope. The ascent is difficult, especially when carrying spears and other equipment. But once at the top, the warmth of the first rays of the sun makes a brief rest quite pleasant.

The view is spectacular. Mountains covered with forests frame a valley crossed by a winding river. Some deer wander along its banks, stopping for a drink now and again. The grasslands of the plains are still in the shadows, but enormous figures can be seen moving slowly near a clump of trees.

The hunters' eyes are riveted to those colossal animals with their long white tusks. The animals are almost motionless, only their long trunks move rhythmically through the grass and shrubs, lifting bundles of foliage to their mouths. This is a herd of *Elephas antiquus*, one of the largest proboscideans that ever lived. Their long tusks, slightly curved at the end, are as big as tree trunks. These animals can reach heights of over twelve feet. It is very difficult to approach them because, although their sight is not very good, their senses of smell and hearing are excellent. At the slightest sign of danger, they unite to form a circle with the weaker members and the young in the center and the stronger and older members threateningly facing outward. They can reach considerable speeds when charging an aggressor.

It is almost impossible to isolate and kill an adult. It would take not only perfect organization, but also far more hunters and great courage. And it would be useless in any case; the group back at the camp is too small to be able to consume so much meat, and it would be impossible to transport the rest. It is much more practical to kill one of the younger elephants, whose meat is also more tender. The problem is to get past the adults' surveillance. This takes particular cunning and skill.

The group of elephants lumbers away from the trees. Patiently the hunters observe them, studying each individual. The dominant male and the mothers must be identified because they are the ones that will react most violently to the attack.

Three young elephants shuffle alongside their mothers. One was born only recently and is no more than three feet high.

Now the entire herd has left the dense vegetation and is moving toward the river, following in the footsteps of the leader, a huge beast more than twelve feet high with a broken tusk.

The morning breeze carries the grunts of the elephants to the hunters.

The oldest waves his hand to draw the attention of the rest of the group. A deep scar on his face gives him a foreboding look. He points to the places from which the group is to attack. Each member knows what to do.

The strategy is adapted to the geography of the valley. The elephants are to be scared and put to flight, and then forced to enter a narrow corridor between the mountains and a bend in the river where the ground is marshy. The herd will inevitably scatter, making it easier to attack one of the young ones, no longer protected by the adults.

The time has come to split up. After handing their spears over to their companions, five of them briefly salute and silently set off for the forest.

From his vantage point, the oldest hunter watches them agilely cross a ridge covered with vegetation. Under no circumstances must the herd below be alarmed. Later, the five men will have to scare the animals, driving them toward the place where the rest of the group is waiting in ambush. The wind is in their favor and the shade lying over the ridge allows them to move unnoticed.

When the five scouts start to descend toward the river directly behind the herd of elephants, the oldest hunter signals to the rest of the group made up of the more expert spear-throwers. It is now their turn.

They also start their descent toward the river, carrying the heavy and cumbersome spears.

Huge boulders that have broken away from the rock wall overhead are scattered throughout the forest and heaped together in the valley near the river. This is where the hunters have chosen to hide.

The muddy ground extending to the river is covered with green grass. When the elephants come stampeding through this corridor, no more than 90 feet wide, the attack will take place, like last year.

The five scouts hidden on the other side of the plain can see the spears poking out from behind the boulders. Everything is going according to plan. Their companions are taking up their positions. They can hear the grunts of the pachyderms and even the sound of the grass and branches being torn away by their long trunks.

A spear is raised and waved slowly. That's the signal. Without a moment's hesitation, the five jump from their hiding places and run toward the elephants, screaming, yelling, and brandishing large branches with which they strike rocks, bushes, trees, and the ground. They also throw large stones.

The elephants are taken totally by surprise. The older and stronger ones hurl themselves at the aggressors with furious roars, raising their enormous tusks. With their trunks flailing back and forth like whips, they crash through trees, bushes, and shrubs. The leader attempts to charge, barely missing the scouts, who are too agile and swift and slip away only to reappear elsewhere. The herd is soon divided: the stronger beasts make a stand, while the others lunge for safer ground—the narrow corridor between the rocks and the water. The elephants plunge down the passage, their roars mingling with deep wheezing and whistling. The hunters in the ambush spring into action.

As the heavy animals start to sink into the mud, their once compact formation falls apart. Some animals can barely make any headway at all, others manage to gain firmer ground and escape. In only a few seconds, the targets are isolated. The newborn elephant has come to a halt and is waiting for its mother who struggles to move forward. But it is too small. The other two young elephants, considerably larger, are pushing ahead, abandoning their mothers who falter in the mud. Perhaps they are already driven by that instinct which will make them increasingly independent in the herd. And that may be fatal.

The oldest hunter, who has not taken his expert eyes off the two young elephants, now decides which one is to be the victim. A fraction of a second later, a long shadow whistles through the air, followed by others, four in all.

The first spear strikes the young elephant full in the throat. The animal is more surprised than hurt. Two more spears fall short, but a fourth penetrates between the hips and the ribs. The skin is still not thick enough to repel the hardened points of these weapons.

The elephant stops, raises its trunk, and lets out a desperate cry. The mother responds with a violent roar and makes an all-out attempt to reach her offspring. But it is too late. Three more spears penetrate the young elephant's rib cage. It wavers for a moment before collapsing onto its hind legs. Two more spears pierce its side. By the time the mother arrives, the young elephant has fallen to the ground. With a roar, she lunges furiously at the hunters, but they have already returned to their refuge behind the rocks.

The mother uproots a young tree with her trunk and hurls it at the boulders, grunting and roaring. Two other females come to nuzzle the young wounded elephant. The enormous males that were defending the rear, including the one with the broken tusk, finally come on the scene. Uncertainty and confusion reign. Some start to charge, others roar. But to no avail. The hunters have disappeared, withdrawing in the same way that they came. Now they just have to wait it out. They observe the scene from their forest hideout. They can see the lifeless body of the young elephant being nudged by the other females.

Some spears have been trampled in rage. But it's all over now. The oldest hunter knows it and so does the huge elephant with its broken tusk. The sun is now high in the sky. The leader of the herd decides to move away from that dangerous spot.

In a chorus of roars, the herd abandons the lifeless body of the young elephant, leaving the muddy terrain pockmarked in the wake of its chaotic flight.

41. ". . . With a roar, she lunges furiously at the hunters, but they have already returned to their refuge behind the rocks. . . ."

The elephants are already far away before the oldest hunter comes out from behind the rocks and walks across the clearing. It looks like a battlefield. The others soon follow. Tradition demands that the elder approach the prey first to make sure that it is actually dead. The others stand by, ready to come to his aid. But it is unnecessary.

One of the young men bends down to pick up a spear. Despite its thickness and weight, it has been snapped like a twig by the elephants. Now he can understand why so much time and care is dedicated to the choice of the wood and the process of hardening it by the fire. An elephant hunt takes a lot of preparation; deer, ox, and wild boar hunts are much simpler.

The five scouts reappear. They are satisfied and smiling. One of them is holding by the ears a rabbit that he managed to take by surprise along the way. It is not unusual prey and constitutes an almost daily element of the hominids' diet.

One of the hunters opens the elephant's mouth. It still has its baby teeth. While the spears are extracted and the animal's body is rolled over for quartering, the stone tools prepared the evening before are pulled out of a pouch. Someone chips at a piece of flintstone found in the area: its sharp edges are excellent for the task at hand.

The hand axes are carefully driven into the animal's skin. Despite its youth, its skin is already very tough, although not as thick as that of its parents.

The cutting edge of the tools has to be honed continually by further chipping. A precise cut opens the abdominal cavity. The innards are pulled out and stuffed into some bags. They are still warm and, once cooked, will provide a delicacy for the whole camp. Some succulent pieces are immediately cut and distributed among the hunters as a "toast" to the kill.

The little elephant is skinned with care. The fat under the skin is important in the diet of these hunters. Practically nothing is left behind. The animal is quartered, its ribs cut into pieces and skewered onto the spears, ready to be transported.

Even part of the head is taken back to the camp. When the group sets off, all that is left on the muddy ground is a shapeless carcass, a few broken spears, and a pool of blood.

Climbing the slope, the hunters slowly disappear into the forest. All are carrying something, either a limb or a heavy bag or the mutilated head of the young elephant, its trunk dangling down behind.

The hunt is over.

The Hunting Life

Organized hunting had already been practiced for a long time. *Homo erectus*, the main character of the story just described, was, in fact, a formidable hunter, as numerous finds from all over Africa, Europe, and Asia demonstrate.

Although our story is imaginary, much evidence leads us to believe that an elephant hunt like the one described could actually have taken place. Among other things, remains of young elephants (fragments of tooth buds or baby teeth) have been found. It may well be that *erectus* drove large prey into bogs, killing them with their spears once they could no longer get out.

The hunting techniques used by *erectus* for different kinds of prey are not known. The remains found in their sites indicate that they hunted an enormous variety of prey. For example, the following animals have been identified from the remains found in the cave in Arago, France (dating back 450,000 years): mouflon, giant oxen, chamois, musk oxen, wolves, lynx, panthers, foxes, wildcats, reindeer, fallow deer, mountain goats, beavers, rabbits, and hares, to name only a few. Many of these animals are difficult to hunt and required specific techniques. But during their long existence in Europe, *erectus* also hunted much larger animals.

If we were to fly over those ancient lands in a helicopter, we would be able to see herds of bison and wild horses in the flatlands, as well as some large African mammals such as elephants, hippopotami, and rhinoceroses, which inhabited Europe in the warm and long interglacial periods.

What techniques were used to hunt such diverse animals?

Let's start with horses. Horses generally lived in open environments, such as grasslands, in very compact herds. It must have been almost impossible to approach them without being seen; they would have inevitably galloped away, keeping out of range of spears. How could they be caught then? The *erectus* cleverly took advantage of the horses' instinct to flee, scaring then and driving them toward other hunters ready for attack, or into dead ends offering no chance for escape. This is a tactic which many predators, including lions, use regularly.

Some carnivores on the savannah, such as African hunting dogs, use a different hunting technique based on cooperation. Individuals are separated from the pack and the dogs then take turns pursuing them, as in a relay, until they collapse. These animals sometimes force their prey to run in a semicircle and then take turns cutting them off until the moment of the final assault.

This strategy is also suitable for humans because, unlike most other animals of the savannah, human beings are, as mentioned, excellent long-distance runners. They can't run as fast as lions, but they can run for a much longer time. They are marathoners rather than sprinters. This technique is still used by many peoples, among whom the Tarahumara Indians in Mexico. They are capable of pursuing a deer for up to two days until it finally collapses from exhaustion and can be killed. It is not unlikely, therefore, that this technique was used by *Homo erectus*.

Tricks and Traps

Investigators feel that the walls of the cave of Font-de-Gaume in France provide evidence of another important hunting technique: traps. Painted on the wall, almost like a snapshot by some anonymous observer, is a mammoth inside a five-sided figure with an open vertex. Many have interpreted this to be a large pit covered with branches and leaves into which the mammoth was driven by the hunters (Cro-Magnon in this case, much more recent).

But a huge pit is not needed to capture large prey. In Zaire, for example, in Virunga Park, poachers capture enormous hippopotami by digging very shallow holes (twelve to sixteen inches wide and twelve inches deep) along the path used by the animals when they leave the water in the evening to return to graze on the savannah. Covered with leaves and branches, the holes are practically invisible and are just deep enough to injure the heavy mammals' limbs, making them easy prey.

The Pygmies also have ways of capturing large animals like forest elephants, surrounding them and repeatedly striking from various sides. Sometimes a single hunter hidden in the vegetation (covered in manure as an olfactory disguise) may courageously drive a spear into the belly of an elephant and wait until the loss of blood weakens and kills the animal (in this case, the spear generally has a poisoned tip).

It is clear that hominids may have had a number of hunting techniques which we cannot even imagine; it is difficult to put ourselves in their place.

One technique, for example, is astounding: catching a hare with the bare hands. The very thought seems impossible (also because we have never tried, since easier techniques now exist). But Louis Leakey, pioneer of the excavations at Olduvai, has demonstrated that it can be done with a little cleverness and a lot of agility, especially when the hare suddenly pops out from under a nearby shrub. Hares typically run zigzag to confuse their pursuers, suddenly swerving from side to side, but they lower their ears an instant before doing so. So, it you throw yourself to one side or the other, like a soccer goalie defending against a penalty kick, you have a 50 percent chance of landing on the beast. If it escapes, Leakey claims, the hare will take shelter and remain immobile in some shrubbery so as to avoid being noticed by predators. But this is not effective with human beings because they not only have excellent eyesight, but can distinguish colors very well and are therefore less fooled by camouflage.

Other situations may have called for other techniques. Witness the personal account of a Japanese man who hid in the jungle of Guam for sixteen years after the end of the last world war and managed to capture various kinds of animals without the use of traps, nets, or spears. He killed wild pigs, for instance, by waiting on the branch of a tree

42. Hares may have been caught barehanded in the past. Louis Leakey showed there was a good probability of success using a special technique to intercept them and then lunging like a soccer goalie.

overhanging a trail where they passed. After a few days, when the animal finally arrived, he would daze it as it ran along the path below him and then kill it by dropping a heavy stone on its head.

The Giants of Prehistory

Many other strategies and tricks may have been adopted and invented by *erectus* to procure animal protein. But it is difficult for us to reconstruct this part of history. The corpus delicti exists, however; that is, animal bones which under the microscope show clear signs of the meat having been stripped off with stone tools. These bones demostrate that hominids living hundreds of thousands of years ago were able, in one way or another, to capture a large number of different kinds of animals, many of which (and this is the most surprising thing) were of gigantic proportions.

Elephas antiquus, for example, the protagonist of the story told in this chapter, was much larger than the elephants that exist today. As stated, it was over twelve feet tall, and had long legs and enormous tusks that touched the ground.

The bison of the steppes, which lived 400,000 years ago, reached heights of almost six feet and lengths of about eight feet. They were like small vans. The Mosbach horse and the Merck rhinoceroses, the remains of which were found among those of the animals hunted around 450,000 years ago by the occupants of the cave in Arago in the Pyrenees, were very large. The horses measured five feet at the withers, while the rhinoceroses (males) were as large as Indian elephants today.

But at the time of *erectus* lived another animal of gigantic proportions, extinct today: the largest primate ever known to have existed, which roamed eastern Asia until just over 500,000 years ago—the *Gigantopithecus*, that is, the "giant ape."

Although only some teeth and mandibles (of awesome proportions) have been found, experts calculate that some specimens were as tall as eleven feet. Others feel that the teeth and jaws are large because of disproportionate growth due to its specialization in chewing, and that the animal was actually smaller (six to nine feet in height or less).

This Asian King Kong was, however, a gentle herbivore, with habits similar to those of the giant panda—a kind of "friendly giant," a formidable plant eater. Its gigantic size did not save it from extinction, however. Some think that *Homo erectus* was the cause of its extinction, as is now the case with the gorilla in Africa and the orangutan in Asia, once present in most of the Far East and now reduced to a few solitary groups in the forests of Borneo and Sumatra.

A few years ago, a group of researchers discovered *Gigantopithecus* remains along with human bones and stone tools in a site in Tham Khuyen, North Vietnam. This has led to the assumption that they were actively hunted by primitive inhabitants of Asia.

The Saber-Toothed Tiger

Other large (and much more fearsome) animals roamed Europe at the time of *erectus*: lions much larger than those of Africa today and the famous saber-toothed tiger. This name has actually been used to encompass a number of genera and species, sometimes quite different. The European species was called *Machairodus* ("Dagger Tooth") and definitely lived 650,000 years ago at the time of *erectus*. Remains of a human mandible and of a *Machairodus* were found in the same layer in Mauer, Germany.

The habits of these ancient carnivores are not known, but it is believed that they hunted mainly large prey such as elephants, bison, and perhaps rhinoceroses, using their long, sharp fangs to slash open the victims' guts and then waiting for them to bleed to death. Or they may have, as lions do today, sunk their teeth deep into the neck, suffocating the prey (or perhaps severing arteries or breaking vertebrae).

The front legs of the saber-toothed tiger were usually much stronger than the hind legs (as in the hyena today), making it a clumsy runner. The beast probably waited in hiding for its prey or crept up to it very quietly without being seen.

Some species of saber-toothed tigers also lived in Italy. Remains have been found in Monte Peglia, Umbria, together with artifacts dating back 500,000 to 700,000 years.

These carnivores lived on the Italian peninsula until almost 100,000 years ago and must have terrified the hominids. Many an unpleasant encounter can be imagined, as those that have taken place (and still take place today) in the African savannah between lions or leopards and natives.

43. The largest primate that ever existed: the *Gigantopithecus*. Some reconstructions hypothesize a height of about eleven feet (according to others, it was slightly larger than a gorilla). Coming upon it for the first time must have been a terrifying experience; but the *erectus* learned to hunt this huge animal. Some people even feel that *erectus* was to blame for its extinction.

However, it would be incorrect to think of the *erectus* as helpless victims. In their own right, they were fearsome predators, the most fearsome of all, as attested by the impressive remains of their kills. It may well be then, as mentioned previously, that animals (including tigers) learned to stay away from these hominids, especially when they were in groups. That should not come as a surprise. On the African savannah today, lions fear the Masai warriors because they know that they can kill them with their lethal spears. And they often do so— each time one of their people or animals is killed. Sometimes merely the presence of a lion that has become too threatening is enough to set off a punitive party. And it should not be forgotten that the Masai are not the excellent hunters that the *erectus* must have been, but mere livestock raisers, who depend on their cows and goats for sustenance.

Hunters but, above All, Gatherers

Thus, *erectus* were great hunters, although probably quite different from the way they appear in certain "bloody" illustrations exalting their almost "gladiatorial" encounters with animals. Yet, despite their expertise, hunting probably only provided a minimal part of their diet. This hypothesis is based on observation of what occurs in many hunter-gatherer populations today.

For example, meat accounts for only 30 percent of the diet of the Pygmies of the Ituri forest and only 30–40 percent of that of the Bushmen of the Kalahari Desert. Strangely enough, the proportion is the same, although they live in quite different environments. Why? The answer is simple—it is not easy to procure meat.

Hunting takes a long time and is often unsuccessful. While waiting for prey, the group still has to eat. That is why roots, berries, nuts, and sprouts are important.

We have no direct evidence of our ancestors' consumption of foodstuffs: while there is a good probability of bones fossilizing, the probability of that happening to roots or berries is almost nil. Therefore, some important data will always be missing from prehistoric sites: how much and what kind of vegetable food these people ate. Nevertheless, a good idea of "prehistoric dishes" can be gained from the habits of hunter-gatherer populations today.

For example, the Bushmen gather a certain kind of nut, the mongongo, that grows in southern Africa. This nut has a high energy content: a few nuts provide the same amount of energy as a large steak.

But the real secret, pointed out by Dr. Marylène Patou of the Institute of Human Paleontology in Paris, is that the nuts gathered in one day provide the energy required for three days, while the energy obtained statistically in one day of hunting is enough for only one day. This is crucial when a certain number of calories is required every day.

Of course, what is true for Africa is not necessarily true for Europe or China. In Africa (despite the alternation of dry and rainy seasons), the tropical climate offers enough resources all year round. But when the *erectus* migrated northward, toward climates with much greater seasonal temperature changes, the availability of fruits and vegetables was not the same. In the winter, hunting may have become more important, also because it provided skins and furs as protection against the cold. Thus, the proportion of calories obtained from vegetable and animal matter varied. An extreme example is offered by Eskimo hunters. During the long northern winter, they eat almost exclusively meat.

In other words, as often happens in nature, each species and each culture develops behaviors adapted to the food supply.

Erectus, however, have left us not only remains of their hunting, hand axes, and hearths. Traces have been found of their huts, which are the most ancient known dwellings. These often very large structures attest to an intense social life, evidenced also by what is known of the furnishings.

Collective life (which will be the subject of the next chapter) calls for a common shelter, above all—a space for the people to come together, warm up, protect themselves, communicate, eat, and sleep.

The next chapter will start with a description of the construction of a hut, based on the evidence available today.

10

Communal Life

The Construction of a Hut

The group of *erectus* sits down on a small grassy clearing at the foot of the rock wall. The climb to this lookout overlooking the valley was the last leg of their long journey.

The more expert hunters, who reached the area slightly ahead of the others, are already looking around. Fragments of the perimeter of the hut built the year before are still visible. A few large stones lying in a row and some stone chips scattered on the ground are all that remain of last year's camp. Someone picks up a biface that is still in good condition. There is no time to lose. The sun is starting to set and the hut must be able to accommodate the whole group for the night.

Two hunters scouting the area return. The group has been camping in this place for a number of seasons already, but a quick look around (for the dens of carnivores, tracks, or edible berries) is always useful. Soon all set to their separate tasks. Some gather wood for the fire: branches, twigs, and larger pieces of dry wood. Others drag longer branches to be used for construction.

Heavy blows shaking the tree tops resound in the woods. With the help of hand axes, three men are cutting fresh boughs for the hut. Some smaller and leafier ones are cut for the covering.

A new member of the group, who is still not familiar with the site, busily drags wood back to the camp. On his way, he runs into women carrying twigs and branches laden with berries. The only people not taking part in this apparent confusion of activity are a pregnant woman and a young man with a recent leg injury, who watch the scene attentively.

One *erectus* starts to survey the ground where the hut stood the year before. Poking around with a stick in a small heap of dark soil, he comes upon some pieces of coal: the old hearth. He immediately understands how the hut was oriented and decides where to plant the corner poles of the new structure.

In the meantime, others have been taking the smaller branches and leaves off some poles that will serve as central supports and are now starting to sharpen the tips.

A lot of chatter accompanies the swift blows. In only a short while, the framework of the hut, composed of the larger poles planted into the ground and some cross branches, is erected. In some ways, it is not much different from modern tents.

A wall of smaller branches and green fronds is gradually erected against the supporting poles. With the help of some large rocks keeping the branches in place, the hut is soon ready.

People continue bringing fronds to cover the structure, adding thickness to the walls. This will keep rain water out and heat in during the night (even though the hut is not draft-proof by any means).

The sun has almost set. A fine wisp of smoke curls from the top of the hut, as some hunters watch the dark shapes of a herd of giant oxen (which will be hunted in the following days) down in the valley. Inside, a fire illuminates the faces of the group.

Everything is ready for the night. The hominids are curled up on pallets made of branches, close to a fire that will burn all night.

Primitive Shelters

Construction of a hut like this seems quite easy and, in fact, many of us experimented with it ourselves as children (albeit probably with more primitive results than *erectus* achieved).

The fragility and precariousness of the construction give an idea of how difficult it must be to identify the remains of a hut after hundreds of thousands of years. Only the positioning of the stones around the poles may last through time. But in the absence of a specific context, even these signs are difficult to interpret.

It is no wonder, then, that there are practically no traces of huts that are more than 400,000 years old. There is, in fact, much debate about other findings dating back over a million and a half years, in particular, a circular pile of rocks about twelve feet wide that was discovered at Olduvai. Not only the position of the stones, but also the geological and sedimentary characteristics of the area raise some doubts and suggest that the pile of rocks is a natural formation.

But it would be surprising if hominids who knew how to work stone—even much earlier than 400,000 years ago—manufacturing hand axes and scrapers, did not know how to put some poles and branches together to build a hut.

Hominids probably knew how to build shelters for themselves from the earliest times, only no traces have been preserved. In any case, the first shelters were probably trees, offering (limited) protection from predators. It may have been while in the trees that the comfort of a pallet was discovered. This is something which chimpanzees are still very fond of: every night they build themselves a rudimental bed of branches and leaves.

Only when the hominids finally started spending more time on the ground did they feel the need for an artificial shelter. Of what kind?

Let's try to put ourselves in the place of a small nomadic community wandering the savannah. What would we do at night if we couldn't climb a tree? We might try to use branches, twigs, and thorns to create a barrier against predators (as the Masai who live in those areas do today)—certainly not insuperable but, nevertheless, a barrier behind which twenty or thirty individuals capable of fending off aggressors (with clubs, stones, etc.) could gather. During the day it could provide a point of reference for the group—both for those out looking for food and for the females staying at home with their young.

True, no large mammals create huts for shelter, though many (e.g., hyenas, African hunting dogs, and foxes) use holes, caves, and natural shelters, or dig dens. But the first hominids had two good reasons for building huts: they had the intelligence to imagine them and the manual ability to construct them.

Unlike a den, a hut is simple to erect and well-suited to the semi-nomadic life of these hunter-gatherers.

Of course, this is only a hypothesis, but it seems rather plausible, considering what is known about the behavior and environment of the early hominids.

These early camps probably developed into real huts—small igloos of intertwined twigs and branches—like those used today by the Pygmies. The hut described at the beginning of the chapter was the final stage of this process. It was produced with the help of the technology of the hand axe, with which *erectus* could cut and shape construction material. And this material permitted the construction of large huts, like those discovered at Terra Amata.

A Cottage by the Sea: Terra Amata

At Terra Amata near Nice, the remains of large huts built at different times have been found at various levels of a site near the sea dating back about 380,000 years.

The dimensions range from 24 to 48 feet in length and from 12 to 21 feet in width. They are almost like small hangars (and bring to mind the large huts used today by some Indios in Amazonia, housing entire communities).

The remains consist of a series of holes in which poles and pickets were planted (sometimes two at a time). These holes clearly outline the perimeter of the hut. Other evidence indicates the cross-section even more clearly: (1) piles of large rocks along the outside perimeter (to keep the leafy boughs on the ground), (2) large central holes for the supporting poles, (3) another large hole probably used as a hearth, and (4) various kinds of remains (bones and chips from tools) inside the dwelling.

Studies of this evidence suggest that the constructions were light and temporary, of the kind described in our account: a makeshift nomad camp for short-term use (it is significant that eleven of these huts have been found, indicating that *erectus* returned to the site—that it was well known).

Studies of the remains have offered some interesting details. First of all, *erectus* probably slept beside the fire. In fact, the area around the hole constituting the hearth is unusually free of debris and leftovers (perhaps it was covered with furs). Furthermore, a small workshop for the manufacture of tools was found in one corner. Stone chips, many of which could be reassembled, were scattered around. A flat stone was located at the center of the workshop. It was probably the seat on which the ancient *erectus* craftsman sat.

Another finding could indicate that hygiene was not a priority among these ancient beings. A "coprolite," that is, a fossil of human excrement, was found inside the habitation. Some people feel that an area may already have been set aside for toilet functions at that time, but in this case, the coprolite was found in the kitchen area. It may have been an accident caused by a child or a sick person. Or simply an oversight. Other coprolites have been found outside of the huts. Analysis, after 400,000 years, has been able to reveal the kind of parasites present in the feces and, therefore, the kind of illnesses from which these hominids suffered.

One last thing has to be mentioned about this site at Terra Amata. It was discovered during construction of a condominium in Nice on the Côte d'Azur. The scientists were given six months in which to carry out all their excavations and surveys before it was cemented over and obliterated. The work was carried out in record time by Henri de Lumley and his team. Approximately 35,000 artifacts were found at various levels. Now, there is a condominium where the huts of the ancient *erectus* once stood. In commemoration of its ancient inhabitants is a small museum—a tribute from the civilization of hot water and elevators to the civilization of the boughs and the coprolites.

A Two-Room Apartment in a Cave: Lazaret

Not far from Terra Amata is the famous cave of Lazaret, also on the Côte d'Azur near Nice, close to the sea.

But we leap forward in time over 250,000 years. The people who lived in this cave were halfway between the *erectus* and the Neanderthals. This is confirmed by examination of the (very few) remains found: two teeth and one skull fragment.

The cave in Lazaret is extremely interesting for the information it provides about the way the hominids who took refuge there 130,000 years ago lived.

It is a natural shelter from rain and bad weather. But it did not offer enough protection from the climate, which was bitterly cold in the winter at that time. The cave was twenty meters above sea level; however, study of ancient pollen has shown that pines grew in the area and that the climate could be compared to that found 3,000 feet above sea level today.

So, for greater comfort in the winter, the hominids built a shelter inside the cave: a hut against the wall (on the right when entering the cave) which was, it seems, covered with skins. This dwelling has been reconstructed from the large quantity of artifacts found on the ground by de Lumley's team. This is what the winter camp must have looked like 130,000 years ago.

The piles of rocks placed around the poles clearly define the size of the hut: 33 by 11½ feet—a space large enough for at least a dozen people. Other rock piles suggest another wall for further protection against the wind.

The interior was divided into two separate rooms. In the more sheltered one, two holes were found which may have served as braziers for embers brought in from the large fire burning in the cave.

Microscopic analysis of the soil has revealed that minute seashells which are usually attached to algae were present around the braziers. This suggests that pallets were made of dry algae. Also the remains of the end bones of the paws of wolves, foxes, and lynx lead to the assumption that furs were used as blankets.

Those animals, along with deer, ibex, horses, rhinoceroses, rabbits, marmots, and giant oxen, were the habitual prey of the Lazaret hunters (remains of these animals were found on the ground).

This and other information provides us with a vivid image of this ancient human group. Two families probably spent the five winter months in that cave, with the men hunting in the area and a fire constantly burning in the cave, where the entire group ate and made tools. Animals were skinned and their furs dried for use as clothes, pallets, and shelters.

In the evening, the group would probably meet inside the inner shelter around the two braziers full of glowing embers. Then they would stretch out on their algae pallets awaiting a new day, a new hunt, and a new season.

But what was the daily life of these early communities like?

It is probable that the communities underwent some change during the course of evolution, but that they all had one basic feature in common: they were hunting-gathering societies. This strongly conditioned some characteristics of the group.

We now have an indirect way of trying to understand social structures of this kind: observation of the last surviving communities of isolated hunter-gatherers. It's rather like trying to reconstruct an unwitnessed event by listening to the testimony of people who experienced (or are still experiencing) the same (or a similar) thing.

What can we deduce from the observations made of these groups?

A Group Portrait

First of all, the groups are always composed of a few dozen individuals, no more and no less: usually from twenty to thirty (sometimes as many as fifty), depending on the environment and the available resources.

In some cases, the group may be a bit larger (as is the case among the Australian aborigines living on the coast) or a bit smaller (as is the case of Eskimos in the Arctic), but those are exceptions.

Therefore, it is likely that the hominid communities were also of this size. It is thought that *erectus* lived in communities of twenty individuals (as the characteristics of certain settlements, such as those at Terra Amata, would seem to indicate).

The density of the group was regulated by various factors, among them the resources available in the area. If we were to try to gather berries and tubers and to hunt animals (every day) for a group of twenty people, we would soon find out that it takes a vast territory to satisfy the continuous search for new food supplies.

Studies seem to indicate that three to four and one-half square miles per individual are needed for survival. This is obviously an approximate figure, which varies depending on the resources available in the territory and the ability to use them. In our society, (agricultural and industrial) technology has made it possible for thousands of people to live where only a single individual could formerly have survived. The tools, organization, and cultural development of those ancient, primordial tribes were much less efficient.

Their rudimental organizational ability nevertheless achieved some important results, especially in the division of labor between males and females, and in group cooperation. The same division of labor is still observed in all hunter-gatherer communities: the women gather, the men hunt. The two activities complement each other.

Anything providing energy may be gathered: not only fruit, nuts, berries, and tubers but also larvae and even insects, lizards, and mice. Hunting mainly involves the capture of prey (rabbits, gazelles, ostriches, buffalo, etc.), but also the preparation of weapons and skins.

However, as we saw in the preceding chapter, hunting was a precarious activity, which provided no security, while gathering offered a reliable nutritional basis. Study of the Bushmen today shows that gathering supplies 70 percent of their diet and requires only twelve hours per week of work by the women. The 30 percent supplied by hunting, on the other hand, requires twenty-one hours of work by the men.

Gathering (and also the need to transport small reserves from one site to another) led to the invention of containers: from simple leaves to animal guts and skins. Much later, this led to the weaving of vegetable fibers and leaves into nets and baskets.

Was There a Hierarchy?

Who gave orders in these ancient *erectus* tribes? Was there a leader or a dominant individual? What can be learned from observing the groups of hunter-gatherers that exist today?

Among the Australian aborigines, for example, the eldest plays a dominant role in the group. The Kung in South Africa, on the other hand, have no real leader; the group makes decisions and settles controversies. Among the Eskimo, the leader is the most charismatic or outstanding person. There is no leader among the Pygmies, either, but respect is dependent upon behavior and ability.

In other words, in a small group (unlike the case of a large society or nation), there are direct and very effective controls on the individual to whom decisions may be delegated. All individuals (both men and women) have the opportunity to formulate and express their opinions.

These small groups must not be seen as rigid and compact, but as constantly evolving. Their composition changes, like a bus with people getting on and off, not only as a result of birth and death, but also through intermarriage with other groups.

The incest taboo is deeply rooted in all human communities. Familial bonds inhibit the sexual drive. In nomadic tribes today, encounters regularly lead to "external" marriages (exogamy) which provide genetic regeneration through an exchange of young blood. At times, individuals change groups, creating crossbreeding among the various communities.

Is it possible that all this took place in the groups of *erectus?* It seems reasonable to believe so. The incest taboo is a very ancient behavior, already present in many primates (macaques and rhesus monkeys, and especially in chimps). As regards bringing in new blood, it can also be seen among pygmie chimpanzees and gorilla groups. Therefore, it is quite likely that tribes of *erectus* had occasional contacts and that some cross-marriages took place. This would also explain a certain uniformity in features and culture—techniques for manufacturing tools and the use of fire, perhaps even the organization of huts. The life of the *erectus* (as that of other hominids) was certainly not very varied, even though every day brought a new and unpredictable adventure.

But they also had a lot of spare time. Calculations show that there was much more free time in these hunter-gatherer societies than in later agricultural and industrial societies. That means that the *erectus* did not have to work tirelessly to organize their lives. The price they paid, however, was less security, less comfortable living conditions, a very short life expectancy, and the constant concern about resources.

A Bushman, asked what he felt after changing from a nomadic to a sedentary life, offered an illuminating response. Pointing to a cow, he said, "Each day when I wake up, I see my food standing before me. That makes me feel content."

Thus, the move to tending herds meant "having your steak in front of you rather than having to chase it," just as the move to farming meant "seeing the berries in front of you" and accumulating enough grain in your graineries to satisfy your hunger in the future as well. But this security called for more work.

It is difficult for us to imagine what *erectus* did in their free time. We must not think that this time could be or was used for creative activities. Lions also have a lot of free time, as do chimpanzees and baboons. That does not mean that they have a better lifestyle, more interests, or creative possibilities.

It may well be that *erectus* dedicated much time to play (as is the case of young primates and predators). The children in the camp could probably often be seen chasing one another or roughhousing, with the adults often joining in. It is now well known that play is an indispensable training ground for learning, through the simulation of future situations, as well as a means of socialization and the strengthening of bonds between individuals.

So, the *erectus* had considerable leisure time, but not much to do with it.

And life was definitely short.

Appendix 2 contains the results of research and studies done not only on the life span of *erectus* (and hominids in general), but also on the injuries, diseases, and traumas that can be discovered from analysis of bones. The picture is anything but idyllic.

44. Children at play were probably a frequent sight in ancient camps. As in the more evolved mammals and especially predators, play is an important mode of learning through simulation of future situations.

11

Genealogical Trees and Shrubs

Reserve Genes

The remainder of this book will deal with the Neanderthals and the evolution toward modern human beings, but first we will briefly review the developments discussed to this point and attempt to link some thoughts and mechanisms that underlie the study of human evolution.

This chapter, or at least the first part of it, will be rather technical. Even though we will attempt to be as clear and concise as possible, the impatient reader can skip a few pages if necessary.

It may seem surprising that the human species was able to adapt *genetically* several times to the environment and evolve toward increasingly modern forms, that is, change its physical characteristics (gait, cranial capacity, dentition, hand structure, nervous system, and so on) in a relatively brief period of time.

How was all this possible in such a short time?

Actually, it is not a question of time alone. The existence of "neutral traits" is an interesting concept suggested by modern genetics. According to this, our genetic makeup consists not only of the "right" genes individually selected by the environment, but also of other genes ("genetic ballast," which in the opinion of some experts accounts for 95 percent of mutations) that often do not come into play but are transmitted "silently." If there are no negative repercussions, they are tolerated by the system and carry out their functions. A sudden environmental stimulus can call them into action.

This also occurs frequently in life. Oil, for example, or uranium, has always existed. Yet, they only became important when environmental changes (in response to technology) turned them from almost unused elements into precious materials. Another example is in the game of poker. Five different (and unrelated) cards can turn into a winning hand if, by replacing one, you get a flush.

In the genetic history of human beings (as in that of all other species), this has occurred repeatedly. Just as an incomplete flush in a poker hand can be seen as a pre-adaptation to a winning flush, so in biological terms do certain neutral genes represent a pre-adaptation to a new trait, a new response to an environmental stimulus.

The Combination Game

This game of genetic combinations and recombinations is actually much more complex than we might imagine. There are characteristics regulated by more than one gene (polygenic traits) and genes that regulate more than one characteristic (pleiotropic genes).

Selection could not act on all genes at the same time, but only on individuals (or often on groups of individuals). So, selection mainly affected traits that were decisive for survival, favoring winning genetic combinations. The others, if they were not harmful, were tolerated.

Imagine what would happen if an exceptionally fast car were to come off the assembly line, even if its door didn't close properly, its roof leaked, and the wheels were too large. Despite these defects, the car would win all races and other models like it would be in great demand on the market. This would lead to the production of similar models, perhaps even with other defects that would also be tolerated (and which might come in handy at some later time).

It is likely that natural selection was based on certain vital "priorities," overlooking factory defects that were secondary and that could be counterbalanced by different behavior or adaptation (and further corrective selection).

This means that the great diversity of individuals that we see today (not only in human evolution but in all of nature) is the result of an infinity of combinations, an enormous number of mutations that gives life an endless gamut of possible solutions. Circumstances will sooner

or later favor (and adopt) some of the latent characteristics accidentally accumulated.

Genetic Kinship

To get back to the topic at hand, that is, human evolution (and possible
genealogical trees), it is clear that a more exact "reading" of genetics
would provide great insight. A file listing the genetic codes of all the
individuals and species would make it easier to understand their connections and kinship.

This would be relatively easy if a telephone book of the genetic
codes of individuals and species (listing not only present but also past
numbers) existed. But for now we have only scattered fragments (although the field of genetic mapping is developing rapidly). But very interesting information can already be drawn from the little that is known.

The basic idea is that the closer their genetic heritage, the closer
two species. For example, if we look at a horse and a dog, it is easy
to see that they are quite distantly related. Comparing a dog and a
wolf, on the other hand, or a horse and a zebra, it is clear that they
are quite close relatives.

Comparison of human beings with chimpanzees proves that they
are very close genetically (99 percent), closer than horses and zebras.
Human beings are increasingly differentiated from apes as successive
comparisons are made—with gorillas (still very close), orangutans,
gibbons, macaques, and so on.

These studies, carried out by Vincent Sarich and Allan Wilson of
the University of California at Berkeley, can obviously only be done
on living organisms. The problem is that we have no living Australopithecines or *Homo habilis* or *erectus* to find out what "genetic distance" they are from human beings and, thus, to establish their degree
of kinship. (This would make it possible to reconstruct a reliable genealogical tree.)

But there is a way to get around this difficulty (at least in part).
It was discovered some years ago and is very ingenious.

Genetic material, or DNA, provides the mold with which living
matter (essentially protein) is constructed. Protein, on the other hand,
is made up of sequences of amino acids. Study of the composition

of some of these sequences can reveal the kind of mold or genetic makeup to which it corresponds.

Living forms of the past have disappeared, but fossil forms have survived. In theory, fossil bones should no longer contain proteins (because the fossilization process substitutes the original proteins with minerals), but it has been discovered that some organic traces do remain. Traces of collagen, a protein found in connective tissue, can be found; collagen has the property of maintaining its characteristics, even in fossils that are millions of years old.

Therefore, with a little bit of luck, fragments of the telephone book from the past can be found in fossils. And precious genetic information can be drawn from them.

This type of research was carried out by Jerold Lowenstein of the University of California at San Francisco. Although his work was mentioned briefly in chapter 2, the technique used in his research will be described in greater detail here, not only because of its interest per se, but because of the new prospects it may open up.

Grating Fossils

The principle of Lowenstein's work is as follows. If any foreign organic material enters the bloodstream of an individual, the body reacts by producing specific antibodies (that is, antibodies with special characteristics that allow them to "bond" to the foreign substance in order to attack and destroy it).

By injecting human collagen into the blood of a rabbit, Lowenstein obtained a serum containing specific antibodies against human collagen. He then did the same thing with the collegen of various kinds of mammals (injecting it into rabbit blood), obtaining various kinds of anti-collagen sera.

He then grated some fragments of fossil bone and inserted the powder into the test tubes containing the anti-collagen sera (obtained from various kinds of mammals), to see which would have the greatest reaction with the collagen of the fossils. As was to be expected, the closer the species, the stronger the reaction. A police force might be a good analogy. If a special police squad (antibodies) is given the description of a wanted criminal, including height, age, and physical characteristics, the more a person corresponds to that description, the greater the "reaction" by the police squad.

If applied to the remains of the various hominids, this technique could provide valuable information on their kinship (and therefore their genealogical trees), especially since molecular biology is now able to calculate the rate of mutation of proteins—the so-called "molecular clock." It has been discovered that the amino acids of which all proteins are composed change at given rates. This happens because there are some amino acids in a series that are essential and some that are not. The former remain unaltered throughout evolution because their function is essential, while the latter may change. For example, cytochrome C, a protein of the cell with a respiratory function, contains 104 amino acids. Thirty-five have remained unaltered for over a billion years, while the others, not indispensable, have gradually changed. It has been calculated that (compatible) mutations occur in these secondary amino acids every 20 million years on average.

The rate is different for other proteins: for example, every 5.8 million years for hemoglobin, every million years for fibrinopeptides.

Thus, study of the mutations that have taken place in the amino acids can provide cautious indications of their average frequency and rate. And this molecular clock can then be used to go back in time in search of common ancestors in various species.

In this way, Lowenstein came to the conclusion that the human "bifurcation" does not date back to *Ramapithecus*, as was once believed, but is much more recent. In Lowenstein's opinion, the genetic lines of human beings and chimpanzees probably diverged just over 5–7 million years ago.

Trees or Shrubs?

Genetics is a field which is likely to undergo intense development in the future. It is fascinating to see how the investigators into our past manage to extract increasing amounts of information from one artifact.

A century ago, the shape of the bones was the only clue the paleontologist had to go on. Slowly it has become clear that a large quantity of data can also be gleaned from soil, sediments, pollen, microscopic analysis, biochemical analysis, and so forth. Today, genetic analy-

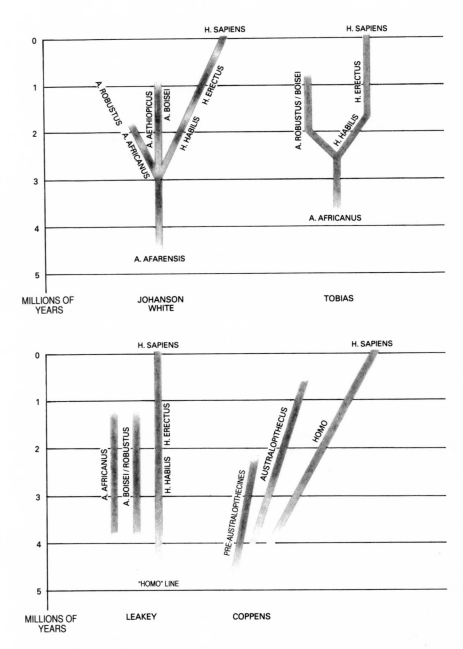

45. Some classic evolutionary trees.

sis has opened a new door. This has one small but all-important draw-back: it requires the (at least partial) destruction of the find. This problem may, however, be overcome in the future, especially if the number of finds increases.

This is, in fact, one of the fundamental problems of human paleontology. An essential requirement for the reconstruction of the mosaic of our past is a large number of pieces. Unfortunately, there are not many field paleontologists, that is, those who work in the field and are directly involved in the initial, fundamental phase of the study of human evolution: the finding of artifacts. This is why genealogical trees are barren and uncertain at the moment.

The opposite page shows some "classic" trees sketched by some of the most authoritative scholars of human evolution. The differences and similarities among these trees are obvious: in particular, all indicate that *Homo habilis* gave birth to *erectus* and *erectus* to *sapiens*.

The only real significant difference is between the tree constructed by Richard Leakey and the one sketched by Donald Johanson and Tim White. Is *Australopithecus* an ancestor of *Homo habilis*? Or did *Homo habilis* descend from a parallel line whose ancestors are not yet known?

Johanson and White claim that *Australopithecus afarensis* is the original stock. Tobias also feels that *Australopithecus* is the common ancestor of all (but in his opinion it is *africanus*, of which *afarensis* is only a variety). Leakey, on the other hand, feels that *afarensis*, like the other Australopithecines, is already a parallel branch and that *Homo habilis* descends from an ancestor of which we have not yet found any trace. Yves Coppens represents an intermediate position (but closer to that of Leakey). He sees Australopithecines converging with *Homo* at some unknown point farther back in time.

We will not enter into the debate, as the data required for a definitive answer are lacking. Perhaps new finds will come to light one day and make it possible to resketch these trees.

We would, however, like to return to a matter already discussed much earlier in the book, and that is the enormous variety (and variability) in evolution, which cannot be reduced to a few labels.

Rather than trees, we feel that evolution can best be described by "shrubs," that is, by an intense ramification of a multitude of tiny branches—a labyrinth in which orientation and the attribution of general names is difficult. Actually, each step of evolution (*Australopithecus, habilis, erectus,* etc.) is a shrub in itself, comprising a myriad of forms and populations, perhaps even species.

We have only a few fragments today of this dense tangle of branches and twigs; it takes only a few simple calculations to show that it is certainly much are complex than we can ever indicate with simple diagrams.

Twenty Billion Skeletons

How many hominids walked the African savannah in prehistoric times? Thousands, millions, or billions? The answer is incredible: from two to twenty billion! Let's do a few calculations.

Suppose that in all of Africa the constant population varied from 10,000 to 100,000 individuals (a rather reasonable figure below which a species risks extinction). Reproductive age was from fifteen to twenty. This means at least five generations per century, or 50,000 to 500,000 individuals per century (as we have seen, life expectancy was also approximately twenty years of age).

If we consider that the period in which traces of hominids have been found spans four million years, this means 40,000 centuries. Multiplying 40,000 centuries by 50,000 or 500,000 individuals per century, we get 2 or 20 billion individuals. That is, from 2 to 20 billion Australopithecines, *Homo habilis, erectus,* etc., may have lived in Africa alone (not taking into account migrations to Europe and Asia). From 2 to 20 times the current population of China!

This figure is rather surprising, especially considering what is left of all these individuals today (or rather, what has been found of their remains). All the fragments of skulls, teeth, jawbones, etc., found up to now would barely cover the floor of a large room.

How can the history of billions of individuals who lived over millions of years in very different places be reconstructed on the basis of such a meager quantity of data?

Actually, an almost miraculous amount of information has already been drawn from those few fossils, thanks to the intelligence and competence of the researchers. It is clear, however, that the reconstruction of a real genealogical tree calls for far more material. It is the scarcity that causes uncertainty and at times controversy over the interpretation of certain fossils.

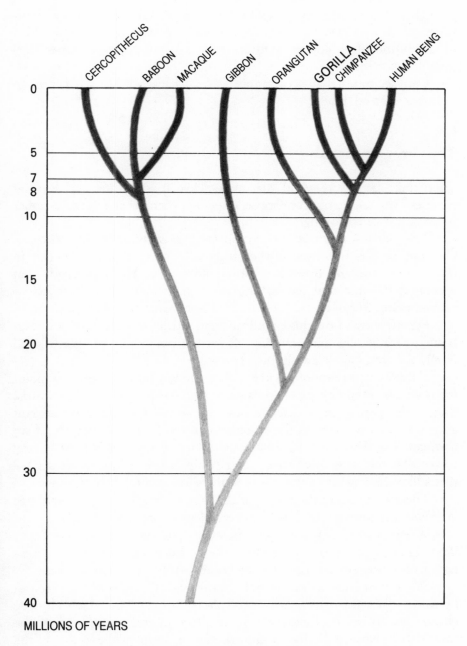

MILLIONS OF YEARS

46. Hypotheses for the evolutionary branching of the primates.

This scarcity also explains why we often read of a single find "overturning" all previous theories. New pieces are constantly being added to the puzzle of evolution: at times they explain things, at times they complicate them.

Can accurate (that is, accurate as far as hypotheses go) genealogical trees ever be drawn?

The Great Crossroads

There may be a way. And that is to leave fossils aside and look at genetics. We are going to propose a new approach that follows a rather unusual line of reasoning.

The starting point is the evolutionary branching of the primates. Genetics proves (confirming the study of fossils) that going back in time means moving toward common ancestors. The crossroads that separated the line that led to chimpanzees from the line that led to human beings should be set at 5 to 7 million years ago, perhaps more.

Fig. 46 shows how this ramification is usually represented: one main branch with smaller ones coming off for the orangutan, the gorilla, and finally the chimpanzee and human beings.

If this representation is correct, then the last bifurcation is the point from which certain characteristics gradually developed. In other words, that is the period in which we have to search for our most distant ancestor, the one born at the beginning of the "human" branch. If we compare Fig. 46 with traditional genealogical trees (Fig. 45), it is easy to see that the latter start much later (only 3.5–4 million years ago)— that's like walking into a movie when the show is already half over.

This is obviously the result of the lack of fossils. Yet, when a tree is drawn, the missing parts must nevertheless be accounted for.

What can be said about this missing part? As we have seen, very little. There is a gap in fossil finds: the remains are few and fragmentary. This whole period has yet to be unearthed. But it nevertheless exists.

We can at least make several hypotheses. The first thing that can be said is that this divergence took place at a time in which climatic change led to the disappearance of the forests and the spread of the savannah in eastern Africa. It can, therefore, be hypothesized that the common ancestor of chimpanzees and human beings started to diversify

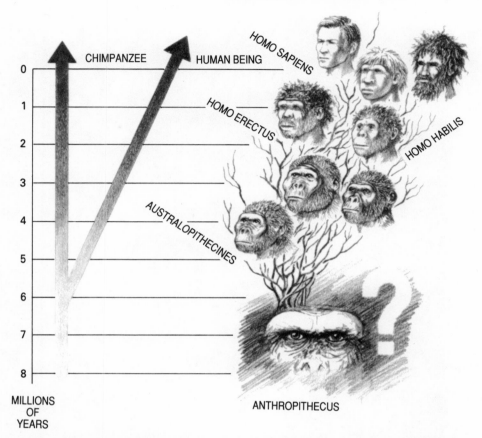

47. It is felt that the branching that led to the chimpanzees on the one hand and human beings on the other dates back 5–7 million years. "Genealogical trees" generally end 3.5–4 million years ago, as remains from the early line of human beings are missing.

at that point, with groups that stayed in the forest (the chimpanzee line) and others that started to adapt to the savannah (the human line).

We don't know how that happened, nor do we know how many or which individuals or groups underwent this adaptation to the savannah.

The research of paleontologists indicates that this profound climatic change, which originated when the Rift Valley was formed through tectonic movements, took place about 10 million years ago. Yves Coppens feels that the newly formed mountains blocked the circulation of

48. A possible face from the past: *Anthropithecus.*

humid air coming from the Atlantic, causing a decrease in rainfall in the eastern part of Africa and, therefore, a climatic change.

Experts in paleofauna tell us that the animals typical of the savannah appeared for the first time between 7 and 10 million years ago.

Looking at a map (Fig. 8), it is easy to see that all chimpanzees live west of the Rift Valley, while all the hominids have been found to the east. Therefore, that is the place (the Rift Valley) and the time at which the evolutionary line that led, millions of years later, to *Australopithecus* and even later to *habilis*, developed.

In other words, beings that had separated from the old branch of the primates and had started the long journey in the savannah toward human conditions lived in that part of Africa for millions of years.

The *Anthropitheci*

Who were these individuals? We have no information about them. But it is reasonable to assume that they initiated the adaptations that were to give birth to the typical characteristics of our ancestors.

These characteristics were probably linked to a new environment —the savannah—which served as a filter, exerting selective pressure.

It is in this little-known period that the ancestors of the various forms of hominids found, dating back 3.6 to 4 million years (with some very rare finds dating further back), lived. They were the grandparents of Lucy, the great-grandparents of *habilis*, the great-great-grandparents of *erectus*.

These pre-hominids, which we would like to call *Anthropitheci* (that is, apes that have taken the path toward the human condition, abandoning the original strain) probably started to differentiate themselves very soon. Or perhaps they already belonged to a group that was undergoing differentiation.

These *Anthropitheci* actually form the stem of the incredible shrub(s) that finally sprouted the uppermost branch: *Homo sapiens sapiens*.

No one can retrace such a labyrinth, but it is very likely that the beginning of branching did not involve too many individuals or groups, and that they were developing new characteristics. With the help of other individuals or groups, they probably gradually built up a new genetic pool, a new species: a new branch heading in a different direction.

If that is how things went (and it is reasonable to assume so), these

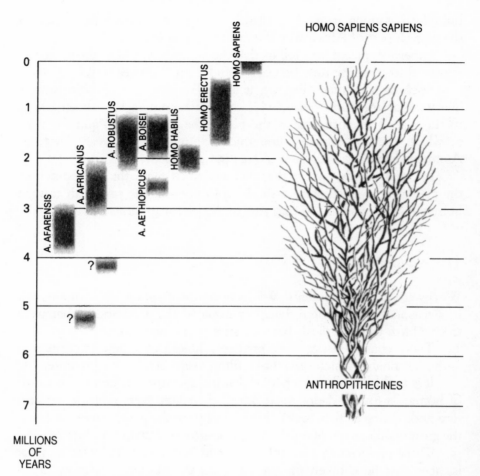

49. An extremely dense and entangled shrub started to grow 5–7 million years ago (that is the period of branching from the chimpanzee). *Homo sapiens sapiens* is only the last (surviving) branch.

Anthropitheci represent the trunk of the shrub of which modern man is the tip (Fig. 49). This shrub has myriad ramifications and branches, representing groups or species, some of which were transformed and some of which went into extinction, giving birth to the only surviving species: *Homo sapiens sapiens*.

Even though we do not know the shape of this shrub, it seems logical to assume that it is rather narrow at the bottom (the birth of

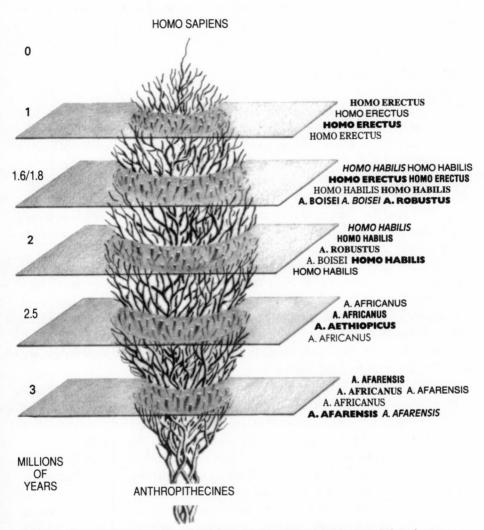

HOMO SAPIENS

0

1

1.6/1.8

2

2.5

3

HOMO ERECTUS
HOMO ERECTUS
HOMO ERECTUS
HOMO ERECTUS

HOMO HABILIS HOMO HABILIS
HOMO ERECTUS HOMO ERECTUS
HOMO HABILIS **HOMO HABILIS**
A. BOISEI *A. BOISEI* **A. ROBUSTUS**

HOMO HABILIS
HOMO HABILIS
A. ROBUSTUS
A. BOISEI **HOMO HABILIS**
HOMO HABILIS

A. AFRICANUS
A. AFRICANUS
A. AETHIOPICUS
A. AFRICANUS

A. AFARENSIS
A. AFRICANUS A. AFARENSIS
A. AFRICANUS
A. AFARENSIS *A. AFARENSIS*

MILLIONS
OF
YEARS

ANTHROPITHECINES

50. Cross sections of the evolutionary "shrub" would probably reveal that there were different individuals in each layer: various forms of Australopithecines, *habilis*, and *erectus* overlapped, diversified, and intertwined in the course of time.

every new genus or species is generally represented that way). And in the case of human beings, it is also narrow at the tip, where *sapiens sapiens* appears.

Shrub-Like Evolution

No one knows what course evolution took between those two extremities, that is, between the *Anthropitheci* and us. We do know, though, that going back from father to father, after approximately 300,000 generations—via a line connecting *erectus*, *habilis*, Australopithecines, etc. —we get to these very ancient ancestors close to the bifurcation from the primates.

Perhaps the shrub can be described as that central line—a trunk with transverse and parallel branches producing individuals sometimes very similar to those on the trunk, but already separate from it.

Unlike shrubs, however, different but close branches can converge in biological evolution (at least until they have not differentiated into different species, which are, by definition, not interfertile). Just as a Watussi can crossbreed with a Pygmie and their children with the children of an Eskimo and a Scot, quite a bit of genetic mixing may have taken place in the past. Not only that, but there may have been evolutionary dead ends, or unfruitful attempts.

All this can be seen by overturning the stones of the past. For example, *Homo habilis* like OH24, 1470, 1813, and OH62 are very different from one another (see chapter 5), but are all crowded into a very short period of time (they all date from 1.7 to 1.9 million years ago).

One of these may well be our direct ancestor, located on the trunk of the shrub. But which one? And which ones are just secondary branches, since none of their descendants led to *sapiens sapiens*?

Fig. 50 shows a shrub with horizontal cross sections at various points. If paleontologists were to be so lucky as to find all the fossils of each of those "chronological layers," they would probably come upon such a large variety of hominids that they would not be able to name them all.

Actually, in order to understand all the details, a cross section of each generation would be required. This would be similar to the technique used by computerized axial tomography (the so-called CAT), in which a series of horizontal sections are juxtapositioned to reconstruct an organ (the same is done in anatomy or in microscopic anatomy, in order to gain a better understanding of the internal structure of an organism by displaying its three-dimensionality).

This general outline of human evolution has been given to show that things are not as simple as they are sometimes made out to be and to warn that certain "trees" are too barren to be able to represent the complexity of evolution. On the other hand, it is also meant as an invitation to those who have the time and the means to do so, to go (or help) look for the remains of those billions of individuals who lived in the past.

That is the only way that we will be able to reconstruct the course of human beings, from Anthropithecines to *sapiens sapiens*.

12

The People of the Cold: The Neanderthals

The Great-Grandchildren of *erectus*

And now on with our story. After the "horizontal" exploration of the last few chapters (which allowed us to examine some important aspects of human evolution, such as brain development, speech, hunting, socialization, and finally, a reconstruction of the course of evolution), let's return to our "vertical" story, that is, the sequence of hominids leading to *Homo sapiens sapiens*.

We never really left this vertical story, since the gradual transformations described in the last chapters document the transitions in time, particularly of *erectus*, throughout the course of a million years.

From the young man in Turkana to the hunters in Terra Amata or Peking, that is, from 1.6 million to 300,00–400,000 years ago, *erectus* dominated the scene. There were significant changes in intelligence and behavior, but oddly enough, the general physical appearance did not change notably from previous periods (as far as one can tell from fossils).

But between 300,000–400,000 and 100,000 years ago something new and fundamental happened in human evolution. There was an acceleration of the evolution of *erectus* toward human forms. In this period, the volume of the brain increased considerably, resulting in beings that were more similar to us.

This took place at the same time in Africa, Europe, and (as far as we can tell) Asia. Why the parallel development? Is it a result of independent local evolutions? Or of migrations? Or a combination of both?

This problem will be dealt with later on. The important thing at this point is that this evolution toward more intelligent individuals manifested itself in two ways: in Europe, with the evolution of *erectus* into the Neanderthals; in Africa, with the triggering of the process that would lead to modern *sapiens sapiens*.

This, then, will be the subject of the last three chapters. In this and the next chapter we will discuss the ascent, triumph, and disappearance of the Neanderthals; in the last, the ascent and relentless progress of modern *sapiens sapiens*.

The Neanderthals did not, of course, appear overnight, like mushrooms. Their dominion in Europe (and elsewhere) from 35,000 to 100,000 years ago was the final stage of a gradual process of transformation. Neanderthals are, in fact, already *sapiens* (but not *sapiens sapiens*), that is, individuals who, by definition, "know" (*sapere* = to know), but do not "know that they know," provided one can make this distinction for individuals about whom we have so little information.

How did this decisive transition from *erectus* to Neanderthal come about? Let's start with the first part of our story and go back 250,000 years to Europe.

The Big Chill

Remains from that time (see chapter 4) point to individuals with evolving characteristics. The German woman from Steinheim and the English woman from Swanscombe already had brain volumes measuring between 1100 and 1300 cc, although their features were still primitive (less accentuated, however, than those of the hominids at Arago, 450,000 years ago, which also had cranial capacities calculated to be around 1150 cc).

The two women, German and English, may have been among the last northern people to inhabit those regions before the big chill: the Riss Glacial Stage (due, like all the others, to cyclical variations in the inclination of the earth's axis, the eccentricity of the earth's orbit around the sun, etc.). This glaciation was not the first (nor was it to be the last), but it was certainly the worst that human beings have ever known. Practically all of northern Europe was buried under snow and ice. Summer temperatures often dropped below freezing; thirty percent of land masses were frozen over. In certain regions of Great Britain, ice reached depths of more than a mile. Forests (even of fir trees) disappeared and gave way to steppes and tundra. The winds changed and icy gusts swept the plains, raising masses of yellowish powder into the air. In short, it was another planet.

Under these conditions, not only the vegetation but also the animals (closely linked to the vegetation) disappeared as a result of either migration or death.

Human beings were also affected by this environmental trauma. They could no longer live in this icy climate and, above all, they could no longer find food to eat. No human fossil remains have been found dating back to the coldest period of this ice age. It is as though human beings disappeared from the scene, either succumbing or migrating elsewhere.

In the regions where life was still possible, on the periphery of the great glaciation, lifestyles changed. The use of fire and skins became more important, as did the ability to hunt in a much poorer and more hostile environment.

The combination of these factors required a greater adaptive ability and, therefore, more intelligence. As Bernard Campbell says, the glaciations were a factor of evolutionary acceleration.

In this setting, after almost 100,000 years of no fossil evidence, forms appear that increasingly resemble the Neanderthals. This trend is already obvious in the late *erectus* (such as the Arago, 450,000 years ago, described in chapter 6), but at a certain point the convergence becomes more accentuated. This is clearly demonstrated by the skulls of the English and German women. In other words, a clear evolutionary line can be drawn from the European *erectus* who were already evolving toward pre-Neanderthals, to classic Neanderthals.

In this chain leading to the classic Neanderthals, two Italian finds at Saccopastore stand out.

Saccopastore: Two 120,000-Year-Old Romans

The accidental discovery of the two skulls at Saccopastore has already been described at the beginning of this book. A worker accidentally drove his pick through one of them while digging in a quarry on the banks of the Aniene near Rome in 1929. The second was found in 1935 by two experts, Alberto Carlo Blanc and Abbot Henri Breuil, while visiting the site. (Four years were required to free it of the concretions in which it was embedded). But what do these two skulls tell us about the subject at hand?

First of all, the brain volume was already considerable: 1200 cc for the former (a woman) and 1300 cc for the latter (a young male), who lived a few thousand years, perhaps five thousand, later.

Actually, it is curious that two individuals who lived thousands of years apart were found in exactly the same place. Perhaps the river had something to do with it: it probably was a very suitable place for a human settlement. Despite the gender differences, the two skulls had similar features: a broad face, a flat head, and a wide and prominent opening for the nose. Both must have had large noses and this was also a characteristic of their successors, the Neanderthals. Another feature they shared with the Neanderthals was the lack of the "canine fossa," that is, the depression (that we have) on our upper jaw between the cheekbone and the canine tooth. This characteristic must have made the anterior part of their cheeks bulge, like someone with wads of cotton on their upper canine teeth (like Marlon Brando in *The Godfather*—only he put them on his lower canines). Their successors, the Neanderthals, also had this bulge.

What distinguished them from the Neanderthals? Smaller teeth and the absence of the protuberance on the back of the skull, the "chignon." Strangely enough, Saccopastore's features are more "modern" than those of the Neanderthals.

Various scholars studying the human evolution of this period believe that the stock to which the Saccopastore people belonged is the same one that later moved to the Middle East and gave rise to a form of Neanderthal quite different from the one that was to emerge in Europe.

Here, too, the evolutionary shrub, the migrations and the "irradiations" from a certain main line, played an important role, giving rise to different individuals in different areas. At this time, there were certain divergences that probably would explain the discovery of different kinds of Neanderthals in Europe and in the Middle East thousands of years later.

One last note about the Saccopastore "couple": the environment in which they lived was no longer that of the terrible Riss Glacial Stage. Studies of paleofauna show that elephants, hippopotami, deer, and rhinoceroses lived in the area at that time.

This was the interglacial stage, that is, a relatively warm period between two extremely cold ones. The terrible Riss was past. Fifty thousand years were to go by before the next great ice age set in: the Würm—the Neanderthal ice age.

Now, let's find out more about this great protagonist of European prehistory.

The Discovery of the Neanderthal Man

In the summer of 1856, a team of workers was digging in a quarry on the banks of the Düssel River (hence the name of the city Düsseldorf). The river runs through a narrow valley at that point, the Neander Valley (*Thal* = valley, *Tal* in modern German).

The name Neander has a curious background: it is the Greek translation of the name Neumann, which in German means "new man." Joachim Neumann, after whom the valley was named, was a local organist and composer of some renown. The valley of the "new man"—a predestined name?

One of the Feldhofer caves lying about twenty meters above the level of the river had to be blown up by the men in the quarry so that they could go on with their work. When they began to dig away the debris, they found some bones, which were simply thrown onto the pile along with the rest.

The owner of the quarry mistook them for bear bones and gave the few that were recovered to a local teacher of natural science, J. C. Fuhlrott. The teacher immediately realized that they were not the bones of a bear but of an ancient human being, a kind of human being never seen before, with a very prominent brow. Probably a victim of the Flood, he thought. He took the bones to an expert, anatomist Hermann Schaafhausen of the University of Bonn.

Schaafhausen confirmed that they were the bones of a human being quite different from us, perhaps a barbarian who lived before our time. He was also struck by the pronounced brow, a characteristic then considered to be exclusive to the anthropomorphous apes.

The bones of the Neander Valley man started to raise considerable controversy. They were the first human remains of the kind available to the scientific community. Actually, similar bones had come to light in the past (in 1829 near Liège, in 1848 near Gibraltar, and in other localities in the eighteenth century), but their importance had not been understood. They were thought to be freaks of nature.

But the Neanderthal find came at a special time. Exactly three years after the find, one of the works that most greatly affected modern scientific thought was published: *On the Origin of Species* by Charles Darwin. If Darwin's theory was correct, that strange individual was none other than one of the links in the evolutionary chain toward modern human beings.

This interpretation was rejected by many. Some people claimed that the differences in the skeleton were merely malformations resulting from rickets, trauma, arthrosis, and arthritis. The eminent anatomist Rudolf Virchow said that it was the skeleton of a "pathological idiot," who must have received a great number of blows to the head during his life.

Someone even maintained that it was a Cossack who, after the battle against Napoleon in 1814, had withdrawn to the cave: the bowed legs were proof of a life on horseback. As for the pronounced brow, it was the result of continuous furrowing owing to pain. . . .

Demonstrating great insight, Darwin and Thomas H. Huxley did not consider the Neanderthal find fundamental, but only one of the most recent variations in human history. Huxley, the great supporter of Darwin, hypothesized that other far more important finds were awaiting discovery in other strata by paleontologists of the future (some of these discoveries have now been made by the people mentioned in this book).

It is odd that the Neanderthals, often considered in the popular mind as extremely ancient beings, are actually very modern individuals, already classified as *sapiens*. In fact, they come almost at the end of our book—in the next-to-last chapter.

A New Misunderstanding

The finding in the Neander Valley was very important not only because it was an excellent artifact, but also because it came at a significant time in the history of science. It brought the problem of human evolution into the limelight.

In the years thereafter, the theory of the "pathological idiot" found increasingly less support as other finds were unearthed (in Belgium,

Moravia, and France). That would have meant a rather disproportionate incidence of "pathological idiots."

The discovery of more and more Neanderthals gave rise to detailed studies. Unfortunately, another error led to a popular misconception which lasted for years. A paleontologist, Professor Marcelin Boule, made an incredible series of mistakes in his analysis of an almost complete Neanderthal skeleton found in a cave at La Chapelle-aux-Saints. Reassembling it in an erroneous manner and not taking into consideration the advanced age of the individual (and, therefore, the malformations caused by arthritis), he presented the scientific community and the public with a being that had a divergent big toe like the apes (and, thus, a rocking gait), was unable to extend its knees, and carried its head forward. He maintained that the being was ape-like not only physically but also mentally, not considering that the cranial capacity was actually greater than ours (1600 cc compared to 1300–1500 cc).

This image of a hunched over, ape-like, and unintelligent being, presented in three prolific volumes between 1911 and 1913, was long to influence public opinion and nonscientific literature.

But things were not really like that at all. We need to take a closer look at these Neanderthals.

What Were the Neanderthals Like?

Let's start with their physical appearance, which we will try to reconstruct from the information available. Reliable data have been accumulated from the many individuals found everywhere in Europe and in many areas of the Middle East.

First of all, height. On the basis of the reconstructions made, the Neanderthals must have been around five feet tall—not much shorter than we are today (in 1800, the Italian population was approximately that height). The build was much stockier, with a much stronger musculature. They would have made champion body-builders. The attachment of the muscles on the bones show that their calves and biceps were highly developed and that they had strong pectoral and dorsal muscles (someone calculated that a Neanderthal weighed twenty pounds more than a modern human of the same height). In the Middle East, the Neanderthals were probably a little taller. The skeletons found measure over five feet tall.

The legs and the forearms were rather short, accentuating a stocky stature.

Set on the powerful neck muscles was a head that was unmistakably designed specifically for that body: large and flat with the maximum width behind the ears, rather like a German helmet. The typical visor-like brow of all previous hominids was no longer straight in the Neanderthal, but arched, rather like an old pair of motorcycle goggles framing the eyes.

Various hypotheses have been advanced about the function of these arched brows, but none are convincing. It is probably a characteristic that has to be seen in the general context of primitive skeletons, which have quite a different architecture from ours. We are struck by these brows since we tend to give great importance to the eyes and everything having to do with them, sometimes overlooking other differences in bone strucuture which may be more important.

It should be pointed out that many people today still have a slight visor, but only in the central part of the forehead, not on the sides.

The most striking thing about the Neanderthals was the size of their skulls. Despite the receding forehead, the average brain volume was greater than ours, approximately 1500 cc (1600 cc in some individuals). The skull had a strange bump at the back, which is called a "chignon" (like the bun worn by women).

Of course, as stated in the chapter on the brain (chapter 7), greater volume does not necessarily mean more intelligence. That depends on the areas that are developed, the wiring, and so forth. But there is no doubt that Neanderthals were very intelligent, as will be seen later in a discussion of the way they lived and hunted.

If we were to see one today, what would strike us about the face, besides the eyes set under very pronounced brows? Probably the large nose, accentuated by the lack of cheekbones. And the bulging of the cheeks over the upper canines (the "cotton wad" effect, mentioned in reference to the Saccopastore people).

Looking at the profile, like that of a mug shot, we would notice the receding chin, the protruding jaw and remarkable shovel-shaped incisors. The long and flattened shape of the head providing for attachment of the strong muscles at the back of the head would also be evident.

51. Research today is reassessing the Neanderthals: no longer the ape-like and primitive creatures once depicted, they are now considered to have been highly evolved, intelligent people very similar to us. The Neanderthals may have been "Nordic" in appearance with fair skin and blue eyes.

Reconstruction of the Face

Some people have tried to reconstruct the face of the Neanderthal man from the information provided by their bones. Their most surprising feature was their enormous nose. The skulls not only have a much larger nasal opening than we do, but the bony root of the nose between the eyes also protrudes much farther forward.

No one really knows what purpose such a large nose served. It has been hypothesized that in such a cold climate as the one in which the Neanderthals lived, a large nose could have acted as a compensation chamber between hot and cold, humid and dry.

Otherwise, it might simply have been one of those neutral traits spoken about in the previous chapter (that is, "hitchhiker" genes, as someone has called them, that get a free genetic ride).

The shovel-shaped incisors were in contrast with the molars, which were relatively much smaller. Both were quite different from the huge millstones of the past: a clear, eloquent sign that the Neanderthals were hunters. Their teeth were certainly specialized not in grinding seeds and leaves, but rather in chewing food composed to a large extent of meat. They had no caries, but a lack of vitamins that depleted the bones might have caused many teeth to fall out.

What is striking about the structure of their teeth is that both dental arches were set well forward, creating the typical protruding jaw mentioned previously. This form might have facilitated the use of the jaws as pincers.

The Neanderthals probably used these pincers not only for biting and tearing food, but also as a vice when working. Their jaws have been defined as a "third hand" for holding objects while working. Just as we may use our teeth to hang on to a corner of our sheets while folding them, the Neanderthals used their teeth to hold meat while cutting it. They also used their teeth while preparing skins. It is believed that the Neanderthals (like the Eskimos and other peoples today) chewed skins at length to soften them for use as clothing in those cold climates. The teeth of some individuals were completely worn down.

The portrait is now almost complete.

The hands and the arms were extremely strong. A Neanderthal would be able to crush our hand in a handshake. Fossils show that their fingertips must have been broad and their nails large. Their hands were suited for relatively unrefined instruments, which were enhanced by the power of their grip. The articulation of the shoulder also attests to the Neanderthals' strength. They must have been strong spear throwers.

Little is known about their gait, but according to Professor Yoel Rak, the cavity housing the articulation of the femur was set further back in certain finds in Israel. This would suggest a different posture and a different way of walking.

But the most intriguing question concerning the Neanderthals is the color of their skin and their eyes. Were they fair-haired and blue-eyed?

Blond with Blue Eyes?

We have not yet spoken about the color of the skin, the eyes, and the hair of the hominids in prehistory. And for one very precise reason: nothing is known about it.

Unlike bones, skin, hair, and eyes do not fossilize. So there is no way of knowing what color they were.

All we can do is look at what happens in nature and note that all populations originating in the equatorial or tropical regions have dark skins. Sometimes very dark. The reason is well known and we experience it ourselves when we tan: the body reacts to the rays of the sun by producing melanin, a protective pigment of the skin. In our case, the tanning is limited to the time that we are exposed to the sun; in the African populations, on the other hand, this is a permanent (genetic) characteristic, the result of a natural selection that has favored those individuals with a more accentuated natural protection (that is, a more pigmented and, therefore, darker skin).

But another important factor affects the color of the skin: vitamin D. The body needs this vitamin to live and the ultraviolet rays of the sun make synthesis of this vitamin possible in the body.

Thus, these two factors have to be taken into consideration: too much ultraviolet light can cause harm to the skin (eczema, burns, tumors, etc.), while too little does not allow for the synthesis of vitamin D, leading to rickets.

The quantity of pigment in the skin balances the equation. In places where there is a lot of sunshine, the pigmentation of the skin provides protection and the abundance of ultraviolet light allows for synthesis of vitamin D. In places where there is little sun, as in northern Europe, it is better for the skin to have little pigmentation: not only is less protection necessary, but the skin has to let enough ultraviolet light through to synthesize vitamin D.

Thus, it is reasonable to think that as the populations gradually migrated northward, those individuals who genetically produced less melanin were favored. In other words, the farther north, the less melanin was necessary. It could, in fact, be dangerous and cause rickets in children.

Taking this into consideration, it is not unreasonable to think that Nordic populations (present and past) were fair-skinned. This process probably started with the northward migrations of *erectus* and continued (or started again) later with their successors, who settled permanently in the north.

It is very likely that the African hominids generally had dark skin (Australopithecines, *Homo habilis*, African *erectus*, etc.) as the primates that live in the equatorial zones basically do (gorillas, chimpanzees, orangutans, gibbons, baboons, etc.) and that the populations like the Neanderthals had fair skin, since they were descendents of people in northern climates. And what about the eyes and the hair?

The issues are closely related. In fact, the color of the eyes and the hair depends on the quantity of melanin genetically produced. Those who have dark skin tend to have dark eyes and hair, while those who have fair skin, tend to have blue eyes and fair hair.

The iris, largely composed of muscle fiber, has many layers. The quantity of melanin in the various layers determines the color of the eyes. For example, those who have pigment only in the deepest layer have blue eyes. Melanin is actually black, but an optical effect makes it appear blue.

Melanin in the more superficial layers leads to other eye colors, from dark black to green to grey. The less melanin, the fairer the eyes. The sum of the coloring of the various layers produces the variety of colors.

Melanin also determines hair color. Those who produce little melanin tend to have not only fair skin and blue eyes, but also fair hair. Genetic crossbreeding creates variations. The typical Nordic individual, with fair skin, blue eyes, and blond hair, is the result of this evolution.

Did the Neanderthal have blue eyes and blond hair, then? There is no way of knowing. But it is reasonable to assume that at least some did (others lived in the southern regions of Europe). And that's not all. The variations and gradualness of this mechanism suggest that some groups of populations (or individuals) had different characteristics: red hair, light and dark brown hair, and different colored eyes. (It may well be, for example, that the Middle Eastern Neanderthals were different because they had acclimatized to different environments.)

The Transience of Races

If the foregoing is true, it means that the concept of "race" is extremely vague. Aside from the fact that we tend to look only at external parameters (such as the color of the skin) without considering other less visible ones (blood group, metabolism, antibodies, etc.), the study of human history seems to indicate, as we will see in the next chapter, that the Swedes are descendants of the Africans, and that the so-called different races blend into one another depending on climatic and environmental situations.

It is quite likely that numerous "races" were born, died out, and were built up again a number of times throughout the course of time, following the migrations and the climatic variations.

Of course, there is nothing to keep people with dark skin, hair, and eyes from adapting to life in a less sunny region. In fact, the Eskimos, who live in the Arctic, do not look at all "Nordic."

But there's a trick here. The body can obtain the vitamin D it requires not only from exposure to the sun, but also from certain food, such as fish liver oil (and the Eskimos are great fish eaters) and milk. But only fishermen or farmers would have been able to take in vitamin D in this manner and, as far as we know, there is no proof that the Neanderthals were either expert fishermen or cattle raiders— and they certainly did not take a tablespoon of cod liver oil every day (as some children today do).

In other words, technology has eliminated many of the mechanisms of natural selection. A white person can live in the tropics and a black person in the northern countries without particular problems of adaptation.

This is also true, to a certain extent, for the American Indians,

52. Preparing skins was very important for the Neanderthals who lived in cold regions. Microscopic examination of their teeth shows that they were often used as a "third hand" to hold skins taut.

who show a certain homogeneity, having occupied the American continent from north to south in relatively recent times.

A blond Neanderthal with blond or red hair and blue eyes contrasts somewhat with the traditional picture painted of this European ancestor. They probably didn't even carry clubs, but much more sophisticated weapons. And they were good tailors, too.

Hides and Hair

Actually, no needles have ever been found in Neanderthal sites. The first bone needles appear with the Cro-Magnon.

But Neanderthals probably made clothing from animal skins, as their worn-down teeth and, above all, the incredible abundance of scrapers seem to indicate. It is true that scrapers could be used for a number of purposes, but it is also true that warm clothing was essential in those cold climates. And, therefore, it was necessary to know how to work hides.

The natural protection of the skin was totally insufficient. At that time, the hominids had certainly already lost most of the hair which earlier ancestors had been covered with. We have no precise data on this matter. In the drawings illustrating this book, the hominids at various stages are shown with less and less hair. That is probably what happened, but it must be made clear that this kind of evolution does not correspond to an increase in "humanity"—not only because the real evolution toward human beings lies in the increasing complexity of the brain (and therefore of behavior), but also because we have exactly the same number of hair follicles as the chimpanzees. The hair on our bodies is just much finer and shorter, so much so as to seem absent in some parts, but an accurate examination reveals this to be an erroneous assumption. We have very fine hair all over our bodies.

Of course, there are also individuals who are quite "hairy"—as a walk along the beach in the summer will prove! It is not that they have more hair, it's just that each hair is thicker.

The brief description given above suggests that the Neanderthals were extremely "new" individuals with respect to the others encountered in this book. And above all, different—products of a very special evolution. Why this relatively rapid specialization?

As Professor Giacomo Giacobini of the University of Turin, one of the foremost experts in Neanderthal populations, says:

> The territory of Europe was unique. It was basically a cul-de-sac, with an obligatory entrance to the east. Internal movements were limited by areas that were covered with ice or unpracticable. This bottleneck to the east created a kind of "island" on which a small population lived. A population more subject, therefore, to selection and to a distinct evolution.

The Neanderthals lived for a relatively brief period of time, almost 50,000 years. Far less than any other kind of previous hominid. And yet we know far more about them.

Yes, we have many Neanderthal remains: we know of 200–250 individuals (some represented simply by a tooth, but others by an almost complete skeleton). That is a lot. Calculating 20 years for each generation, that means one individual for every twelve generations. Prior to the Neanderthals, we had only two examples of skeletons: that of Lucy and that of the young *erectus* 15,000 of Lake Turkana. The greater number of Neanderthals is, of course, a result of their better preservation which has been linked to their custom of burying their dead.

But what is known about the lifestyle of these people who left traces of their hunts, tools, food, and culture?

Following this physical reconstruction of the Neanderthals given in this chapter, let's see whether the remaining information can provide us with a reconstruction of the way in which they lived.

We will start with the background, that is, the environment and the climate, which had such a strong influence on the way they lived.

13

Life (and Death) of the Neanderthals

Between Two Ice Ages

Accustomed as we are to the present temperate climate, it is difficult for us to put ourselves in the place of the Neanderthals and imagine Europe as an icy, windy, hostile expanse.

The geological time chart marking the onset of the glaciations (which can be deduced from the study of the movements of the glaciers) indicates that after the terrible Riss Glacial Stage, which lasted from around 250,000 to 125,000–150,000 years ago, Europe succumbed to another —the last—great ice age which started approximately 80,000–90,000 years ago and lasted for almost 70,000 years, with two brief interglacial periods of warmer temperatures. The first of these took place between 55,000 and 60,000 years ago, the second between 32,000 and 35,000 years ago (these dates are approximate, but rather reliable). This glacial stage is called the Würm, from the name of the valley in which the first studies were carried out. The various periods have been named Würm I, II, and III.

The classic Neanderthals lived between the first two of these periods. They are, in a certain sense, a product of these ice ages.

The environments in which they lived and hunted obviously varied with the latitude, the region, and the period. At the beginning of the Würm Glacial Stage, the vegetation was similar to that found in the highlands today: chestnut, oak, hazelnut, and pine trees. As the cold became more intense, the vegetation thinned and in certain areas gave

221

53. After migration from their ancient place of origin on the African savannah, cultural rather than biological evolution allowed human beings to adapt to a wide variety of climates, even very cold ones as in the case of the Neanderthals.

way to tundra and steppes. Snow and ice advanced over the colder regions, such as northern Europe and the Alpine belt.

In the brief interglacial periods, there was once again lush vegetation with open forests, but the climate continued to be typical of Nordic countries today.

As always, the vicinity of the sea gave the regions bordering on the Mediterranean a more temperate climate. It is quite likely that many Neanderthals headed for these regions, drawn by the milder temperatures. At least, that is why Neanderthals moved to the coasts of northern and central Italy.

Of course, the basic requirement in such a hostile environment is food. The hunting grounds of the Neanderthals can be determined by their prey. What did they hunt in that period and in such a climate? We may be able to use that kind of information to reconstruct the way in which the Neanderthals lived.

Some very interesting information related to food has been patiently pieced together though fieldwork. Analysis of the remains of meals found in Neanderthal caves and sites has made it possible to reconstruct a "menu" from that time, giving an indirect idea of the animals that roamed the territory.

In the cave at Tournal, France, the remains of meals were found mixed with prehistoric implements dating back 50,000 years. Whoever lived in that cave had to deal with intense cold, despite the vicinity of the sea (only nineteen miles away); but survival was, nevertheless, not a major struggle and the inhabitants managed to cook their favorite foods.

What would we have eaten, had we sat down around the fire with the Neanderthals of Tournal?

Studies carried out on the bones found at the site by researchers like Marylène Patou provide rather clear indications of the daily diet of these people. For example, it is obvious from the remains left in the cave that these people were skillful hunters specialized in the killing of horses. Adult animals could provide up to 400 pounds of meat.

Let's sit down and share a bite with our prehistoric ancestors.

At Dinner with the Neanderthals

After a few hours of walking, the group of hunters finally recognizes familiar geographic features and woods. Shrubs with recently cut branches become more frequent. The sound of children's voices and a column of smoke, now clearly perceptible, signal the end of the journey.

When the hunters meet the rest of the group, they put down their loads, and in short, simple phrases tell about the hunt, the prey that got away, and the killing of the horse they have brought home.

Actually, they have brought home only a few pieces. The entire spine was left at the site (too cumbersome, heavy, and unnutritious to bring back). Instead, they exhibit a pile of bloody limbs and ribs and the head with its glazed eyes and gaping mouth.

To the amazement and admiration of all, a powerful hand reaches into a pouch dripping with fresh blood and pulls out a string of intestines and other internal organs. These are considered delicacies (today they are given to the cat!). Someone returns to the cave to stoke the fire in preparation for the meat.

While munching walnuts and hazelnuts gathered in the vicinity, some of the group sharpen flint scrapers and start to carve the horse meat. With expertise, the tendons, joints, and ligaments are cut out. The hide and the long tendons are carefully removed and stored. They will serve for the manufacture of pouches, pallets, clothing, and windbreaks. Of course, the hide will have to be beaten and chewed, in order to make it supple. The tendons will serve as strings.

The embers are now ready, as are the pieces to be cooked—skillfully cut "filets" and chunks with bone.

The cave is actually no more than a large corridor, so the butchering has to be done outside. The floor is covered with the bones and leftovers not only of horses, but also of ibex as well as some remains of boar, giant oxen, bison, deer, and reindeer. Much of this is from previous years, but the bones of the ibex, for example, are recent and give off an unpleasant odor (that we would find unbearable).

As they enter, the members of the group crowd around the fire. There's a lot of excitement. The flames distort the protuding faces with the pronounced and arched brows intent on chewing tubers and berries.

The cave soon fills with the fragrance and smoke of the roasting meat, propped up over the embers on bones or sticks. All eyes watch as the pieces of meat are carefully turned. Flames leap up from the dripping fat.

Chunks of liver (considered the best part of the animal) are cooking on one side of the fire. Before long, someone grabs a piece and, after blowing on it and grimacing because of the heat, pops it in his mouth. There is a general outburst of laughter. Slowly all start to choose the most succulent pieces. No conversation can be heard.

The Neanderthals bite eagerly into the food, using their hands to tear the last morsels from the bone. The meat is still raw inside (rare, so to speak), so this takes some effort. Some help themselves with scrapers. These scrapers are the only real cutlery. The meat is grasped by the protruding and charred bones (rather like eating the drumstick of a chicken). The leftovers—bones that are black at one end and white at the other—are thrown into the recesses of the cave (that is how burned bones were found among the various sediments).

These Neanderthals certainly don't have our concept of hygiene. Nor is their method of cooking very sophisticated. Nevertheless, twigs of aromatic herbs are sometimes thrown onto the fire or stuck into the meat to give it more flavor.

For "dessert," someone smashes a bone to extract the marrow with fingers or tongue. Nuts are brought in for those few who are still hungry. This providential banquet (to be consumed within a few days, before the meat spoils) has sated all. Many stretch out and go to sleep.

Tomorrow is another day and they will start to plan another hunt. Maybe an ibex this time.

Examining the Remains of a Cave

The description of this imaginary dinner is based on what is known of the meals of Neanderthals, that is, the fire sites, the tools, and the leftover bones. Some researchers are still attempting to gain further information.

For example, from the study of the bones of prey found in the cave at Tournal, Dr. Patou has deduced that they provided the hunters with a ton and a half of meat! This is certainly a lot of meat, but upon further consideration, it is not as exorbitant an amount as it may seem at first glance. Given the number of people who probably composed the group, the length of the various stays at the site, and their frequency, Patou feels that steaks were not exactly abundant.

It is logical to assume, therefore (as described in the reconstruction), that the Neanderthals integrated their diet with many vegetable and energy-rich foods like walnuts, hazelnuts, and chestnuts.

While fats and proteins were obtained from meat and some vegetables, the only source of sugar for the hunters were fruit and tubers (of which there are no fossil traces). It must not have been easy to procure this kind of food (especially in certain seasons of the year) in territories affected by the glaciations. But this problem could have been solved in part by the seasonal moves of these nomadic hunters.

It is clear, in fact, that the cave was frequented each year during very specific and limited periods.

Many researchers feel that there was a scarcity of plant matter available in Europe during the cold periods of the Würm Glacial Stage and that this called for more meat than vegetables (the foods that dominate in the more temperate and tropical zones) on the menu, especially in the northern regions.

This kind of diet is not incompatible with human habits. The Eskimos and other populations eat so much meat and animal fat that it would make any nutritionist's hair stand on end.

But they have adapted physiologically. Perhaps the Neanderthals did, too, during the cold periods.

Hunting in Ice-Covered Europe

Another excellent adaptation of the Neanderthals to those places and climates was the variety and efficiency of their hunting techniques.

The bones found among the leftovers of their meals prove that they must have had admirable weapons and strategies. Excavation of the sites has provided some interesting data.

First of all, the wildlife changed considerably as a result of the numerous climatic fluctuations (and the consequent environmental changes throughout the 50,000–70,000 years in which the Neanderthals existed). During the periods of most intense cold, the steppes were inhabited by horses, reindeer, musk oxen, mammoths, bison, woolly rhinoceroses, arctic foxes, and ibex. During the warmer periods, as forests reemerged, deer, boar, megacerous animals, roe deer, beavers, and many kinds of cattle (there are even fossil remains of hippopotami in the initial phases of the Würm Glacial Stage) returned in abundance.

The need to change hunting strategies and lifestyles through the course of the millennia was skillfully resolved by these exceptional hunters, able to conquer such formidable beasts as lions, lynx, wolves, and bears—probably for their furs.

This seems to be an appropriate place to dispel the widespread and erroneous belief that the Neanderthals hunted cave bears.

Since both lived in the same environments (caves), there is a deep-rooted conviction that they were in perennial conflict and that bear hunting had symbolic significance—indeed, that it was almost a rite—for the Neanderthals.

This belief arose from the enormous number of bear (*Ursus spaeleus*) bones found in several caves inhabited by the Neanderthals, sometimes mixed with leftovers from meals. The assumption was that the bears were prey.

Actually, the bears (like hyenas and lions) occupied the caves during hibernation when the hunters had already left on their seasonal wanderings.

And they often died there. Their bones inevitably ended up among the remains left by the Neanderthals, sometimes even in successive strata, giving the impression of contemporaneity. Actually, on an attentive examination of all the sites, only in one case were the remains definitely identified as being those of a bear that had been hunted, killed, and eaten by the Neanderthals. That is not enough evidence to prove that this gigantic omnivore, which measured some nine feet when standing on its hind legs, was systematically hunted.

We do not have precise information on the hunting strategies of the Neanderthals, but some may be ruled out. They did not, for example, drive herds of wild horses off precipices (as hominids from a previous period in Solutré were thought to do). If that had been the case, the skeletons would bear the signs of numerous bone fractures and these have not been found.

It is likely, though, that the Neanderthals forced the horses to pass through narrow gorges or passages (at the bottom of gorges), so that they could attack them more easily. (This was not possible, however, on the steppe, where the horses were always on the move.)

It is obvious that these hunters had a thorough knowledge of the territories in which they lived. The locations of the sites suggest that they chose them close to sources of raw materials for their tools (flint) and to the migratory routes of their prey. This may have been on the border of different kinds of environments, increasing the potential number of animals to be hunted.

Among the prey were the mammoths and the woolly rhinoceroses, both covered with a heavy brownish coat, as can still be seen on the carcasses of the animals that were found intact in the permafrost (a layer of the ground permanently frozen) in the northern parts of the former Soviet Union and eastern Europe.

How were they hunted? By digging holes and filling them with pointed sticks? Or by driving them into bogs? All hypotheses are equally valid as no fossil traces remain. It should be pointed out, though, that holes can hardly be dug into the permafrost (present in a large part of Europe at the time). Furthermore, the fossil remains seem to indicate that mammoths were not frequently hunted.

There was also a technological follow-up to the hunt. The animals killed had to be skinned and the hides and furs tanned.

The Neanderthals had a wide variety of tools for this purpose, called "Mousterian" (five different kinds have been identified). Microscopic analysis has revealed that these implements (above all, large scrapers) must have been used on a number of different materials: wood, meat, horn, and bone—much as we use jackknives today.

Oddly enough, these tools remained unchanged throughout the history of these people—evidence of their cultural homogeneity and relative isolation. This lack of innovation may be rather embarrassing for ancestors with a brain that was generally larger than ours. Yet, as we will see later, driven as they were by particular pressure and stimuli, the last Neanderthals produced much more sophisticsted tools, almost like those made by modern *sapiens*, demonstrating remarkable ability and versatility.

The Importance of Useless Objects

Some extraordinary objects have been found in Neanderthal sites: useless objects.

Why extraordinary? Because in some way they attest to the birth of an esthetic sense. To pick up (and at some point manufacture) a thing that does not serve to cut, divide, hunt, or perform some other practical purpose, but is merely esthetically pleasing, denotes a change in the brain.

An esthetic sense may already have been developed earlier, but no evidence has ever been found. The only exception is Singi-Talav in India, at a site over 200,000 years old: six quartz crystals (the largest of which measures 2.5 centimeters) of no apparent use have been found. The owner of this tiny "treasure" had carried it at least two miles, which means that it had been picked up intentionally.

True, some birds are also attracted to bright, colored objects, but in the case of human beings, the choice seems to be based on other criteria.

To get back to our Neanderthals, two marine fossils (a gastropod and a madrepore) and two pyrite nodules have been found in a cave at Arcy-sur-Cure in France. These are objects that we might well have picked up on a walk and taken home as curios. And that Neanderthals 40,000 years ago did so, brings them much closer to us than any other of their traits do.

It may also be that the unusual shape of these objects gave them some particular significance or made them instruments of some sort of cult (burial customs had been practiced for thousands of years already).

In Tata, Hungary, a curious rounded pebble with two fine perpendicularly intersecting lines engraved on it was left on the floor of a Neanderthal cave 50,000 years ago.

The First Uses of Color

And then there are the first traces of the possible use of color. The Neanderthals did not leave any paintings behind (the first spectacular rock paintings date back to Cro-Magnon times, 15,000–20,000 years ago), but they may have been familiar with pigments: manganese dioxide (black) and red and yellow ocher. Traces of powder and fragments with signs of scraping have been found.

What these dyes were used for is not known. Since traces of ocher have been found in certain graves, some colors may have been used as body ornaments.

We have absolutely no knowledge of this important aspect of

prehistoric culture. Dancing, for example, is one of the most ancient art forms, as is singing. But no one will ever know whether the Neanderthals engaged in primitive forms of dance or song.

No necklaces, bracelets, or amulets have been found in the graves or at the sites, nor have any teeth or bones perforated for use as pendants been found.

Someone has pointed out that the Neanderthals already possessed instruments suitable for engraving or sculpting bone and stone, but we have no evidence of these activities.

Only 35,000 years later (in a period of transition) do primitive bone instruments bearing notches and symbols appear. These were the last works of the Neanderthals before their extinction.

Speech and Socialization

We know even less about the speech of these people who dominated Europe for about 50,000 years. Several studies aimed at understanding the phonation of the Neanderthals, that is, their ability to articulate sounds, have been carried out on the basis of the remains found.

According to some researchers, the famous skull of the "old toothless man" from La Chapelle-aux-Saints shows that the larynx was too high to be able to articulate certain sounds such as "ee," "oo," and "ah."

Recently, however, the hyoid bone of a Neanderthal was found in a grave in Kebara, Israel. (The hyoid bone is a small horseshoe-shaped bone situated between the base of the tongue and the larynx, providing the point of attachment for eleven muscles.) This bone, fundamental for the articulation of sounds, was studied by Y. Rak and B. Vandermeersch and found to be almost identical in shape and size to that of human beings today. This means that there have been no substantial changes in this structure (nor in the bones of the inner ear, nor, presumably, in those of the larynx) in the last 60,000 years. This would lead to the conclusion that the morphological bases of speech were already developed at the time of the Neanderthal.

On the other hand, as already mentioned in the chapter on the brain, the quality of the message must not be confused with the quality of the "loudspeaker." Speech is, in fact, a product of the brain, rather than of the larynx or other parts of the "loudspeaker."

54. No traces of burial dating beyond 100,000 years ago have been found. This does not mean that ancient humans were indifferent to the death of the members of their families or of their friends; the dead were simply not buried.

The English mathematician Stephen Hawking, a man of great genius suffering from a serious nerve disease, used to speak only in guttural noises (which his assistant brilliantly translated). Now he can "speak" only with the aid of a "talking" computer. Nevertheless, he has been able to communicate his thoughts on nuclear particles and gravitational collapse. A deaf-mute can also communicate very complex thoughts with no "loudspeaker" at all.

Whatever the vocabulary, phonation, and pronunciation of the Neanderthals, their communication was certainly intense, because they had a very active social life.

This social life was also reflected in entirely new relations with other individuals of the group. In no previous period can such marked signs of socialization be found as in the Neanderthal era.

This "old toothless man" of La Chapelle-aux-Saints is a typical example. The skull has only two teeth (the two left canines). But if these teeth were already missing during life, as seems to be the case, he would have to have been helped by the others. The skeleton also shows a number of other ailments (arthritis) which would have required further assistance and aid. It is unlikely that he would have survived in any previous society.

There are also evident signs of handicapped individuals in Shanidar in Iraq. One skeleton of a forty-year-old indicates that the individual must have been blind in one eye and unable to use the right arm (due either to an accident or a birth defect). Here, too, the individual was probably assisted by the community, as survival must have been difficult without the use of an arm.

But the most telling sign of group ties is an unprecedented custom: the burial of the dead. The Neanderthals were the first people to take care of their cadavers and perhaps to try to give significance to death. (Today, the graves at Qafzeh, which will be discussed later, seem more ancient, but they belong to more modern beings: the *sapiens sapiens*.)

Burying the Dead

This cultural milestone, which was to lead to new and unforeseeable developments in the following millennia, is worthy of comment. In certain civilizations (such as that of ancient Egypt), care for the dead came to represent the focus of life itself.

However, burial of the dead is not as widespread a practice as one might think. Today, many populations simply abandon their dead to the predation of carnivores. However, it would be wrong to consider any culture that does not bury its dead as "primitive"; it simply has different customs.

At times, these customs are of importance for survival. This is the case for the hunter-gatherers of the Kalahari desert: "Having to take a dying or old person along on a journey towards a spring in the middle of the dry season," underlines Phillip Tobias, "may compromise the survival of the entire group. Thus, it is preferable to build a hut, leave them, and come back with water. At times they find them alive, at times the hyenas have got there first."

Even the Masai, a population of herdsmen with remarkable cultural traditions, adopt systems that may seem "cruel." The sick are usually taken from the camp and abandoned in the savannah (sometimes with a young goat tied nearby) at the mercy of nocturnal carnivores such as hyenas and lions, because a death inside the hut would be a bad omen for the entire group.

There is a multitude of other customs having to do with death that do not involve burial—from cremation to the exposure of the body to the elements (perhaps on a canopy or in the trees, as some North American Indians do). In other cases, the bodies are left to scavengers and vultures or thrown off cliffs or wrapped in animal skins and abandoned on the ground (as in some Eskimo groups) if it is too hard to dig a grave.

These are obviously methods that do not allow for the preservation, much less the fossilization, of the bones.

The advantage of burial is that the bones can fossilize (if the soil is suitable) and are "protected" from atmospheric agents. The most ancient Neanderthal graves are 80,000 years old. Since that time, great care has been taken in the preparation of tombs.

In Le Moustier, France, an adolescent was buried on his right side, with his head resting on his arm as if he were asleep. The old man from La Chapelle-aux-Saints was curled into a grave four and one-half feet long dug especially into the hard limestone of the cave only twelve feet from the entrance. Over his head were found large fragments of flat bones and the remains of a bison foot.

Many have interpreted these artifacts as proof of the belief in life after death, like the offerings of food and tools in ancient Egypt.

There are doubts, however, as Bernard Vandermeersch explains: "On the basis of the data available today, it seems difficult to accept these interpretations completely."

Other graves seem more convincing. In a cave in Teshik-Tash in Uzbekistan, a nine-year-old boy was buried with six pairs of ibex horns arranged in a crown around his head.

The cave in Shanidar, Iran (already mentioned) also contained some exceptional finds. During the excavation, carried out in the 1950s by Ralph Solecki (which led to the discovery of seven Neanderthal individuals), analysis of the sediments around the remains of a hunter buried 60,000 years ago revealed an unusual concentration of pollens. These were often in groups, as if they had come from a bouquet of flowers. Studies revealed that most of the blossoms were very colorful —thistles, yarrow, cornflowers, hollyhock, etc.—as if the body had been placed on a bed of flowers.

But more careful study revealed that all the flowers had therapeutic properties. Perhaps the dead person was a kind of witch doctor, buried with the "tools of his trade." We will never know.

Nor will we ever know why certain bodies were buried in the fetal position. In some cases they are so tightly curled up that they may have been tied. Were the living afraid that the dead might come back to life? Or did they have to reduce their size to fit them into smaller graves dug into the rocky ground?

The reason why the skulls of some bodies were removed some time after burial will also remain a mystery. (This happened in Kebara, Israel, and perhaps in Le Regourdou, France.)

Were there special cults connected to burial and to the skulls in particular? And if so, do the isolated skulls that have been found, like the one in the Guattari cave, take on new meaning?

All these hypotheses are plausible, given the intelligence and spirituality of the Neanderthals. Yet, the paucity of artifacts and the absence of proof make it impossible to give a definitive answer. They could have been subsequent acts of vandalism or the work of carnivores.

We can, however, answer other questions concerning burial. There is so much information available that the epilogue of an event that took place around 50,000 years ago in southwestern France (La Ferrassie, Dordogne) can be reconstructed. The description in the following story corresponds exactly to the burial remains found at La Ferrassie.

A Neanderthal Burial

It has just started to rain in the woods skirting the rocky face. Invisible drops fall irregularly from the grey sky, typical of early autumn, onto the yellow leaves hanging from the trees and carpeting the ground.

The silence of the forest (which is very similar to the ones found in northern Europe today) is broken only by the light patter of the rain and the flight of some blackbirds.

Suddenly, another sound can be heard. It is intense and regular, like footsteps in the distance. A group of hunters appears among the tall, dark trees. Some are covered in skins and furs, others are bare-chested, revealing strong pectoral muscles. There are also some women and children.

At first glance, these people do not seem very different from us. Perhaps it is because they are still some distance away, or perhaps because their fair hair softens their harsh features. They are carrying a heavy bundle wrapped in deer skins (an animal that is becoming increasingly rare).

Upon reaching the foot of the rock wall, they survey the ground covered with acorns, leaves, and stones. A small mound of newly heaped soil marks the grave of another member of their group.

Twenty days earlier, one of the oldest hunters (he was over forty) suddenly died of an unknown cause. He had cried out, clutched his hands to his chest, and fallen to the ground with a grimace of pain.

The others had rushed over to revive him, but in vain. It had been some time since he had taken part in the big hunts. But he had helped gather fruits and with his experience had taught the youngsters the techniques of hunting and tool manufacture. Above all, thanks to his charisma, he had been a leader for the whole group when important decisions, like the choice of route or seasonal moves, had had to be taken. For some time now, he had helped tan hides, chewing silently, his smile revealing teeth worn down to the root.

His woman, who had been suffering from a bad cough for some time already, was not able to withstand the sudden loss. Her condition worsened in spite of the care and therapeutic herbs. One morning she did not get up from her pallet. Her children were the first to notice that she coughed no longer.

Now they are watching disconsolately as a few hunters dig a small grave with some wooden picks. The grave is no more than twelve to sixteen inches deep, aligned with that of her companion. As the group's tradition dictates, the two will also be together in the afterlife. That is why the woman is buried with her head beside that of her companion, as if they could speak to one another (but perhaps also to recall the profound tie between them). Aged beyond her thirty years, her strong features have dissolved into a pallor which touches the hearts of those present. Some girls cry silently, tears rolling down their cheeks. But tears are also welling up in the blue eyes of some of the men: the group has lost two of its members.

Soil mixed with gravel slowly covers the body curled up on the deer skin, almost as if it could feel the cold despite its furry covers. While an older member of the group makes propitiatory gestures over the body, another hunter looks around for danger (especially carnivores). His gaze falls upon two barely perceptible mounds not far away.

They are the graves of two five-year-old children who died last year during the unusually cold autumn, just before the group left.

He recalls that during the burial of the two children he had noticed nine other small mounds, set out in three neat rows, close by. At that time, the oldest hunter had explained that many years before the son of one of the ablest hunters had been stillborn, much to the father's chagrin.

A special rite had been performed in an attempt to ensure reincarnation or life after death. The ashes from the fire had been mixed with gravel and nine little mounds (like wombs) had been piled up, to represent the mysteries of the nine lunar cycles that lead to the birth of human beings. The stillborn child had been buried in the last mound, together with three of the most beautiful tools made by the father, perhaps in the hope that the child would become an able hunter as his father would have wished.

The elder had also told him that other individuals (including children and fetuses) had been buried in that place and that there had to be a triangular stone slab somewhere. He vaguely remembers having come upon it once as a child with his companions. There were strange cup-shaped engravings on the underside, of which no one knew the meaning.

They had also dug up a tiny skull (without the jaw) from that site. It was so small that it must have belonged to a child. They had

55. ". . . Soil mixed with gravel slowly covers the body curled up on the deer skin, almost as if it could feel the cold despite its furry covers. . . ."

been so scared that they had hurriedly put everything back as it was and covered it up.

The elder had told him that he did not know who had buried the child, perhaps a previous group with other traditions. He had heard rumors of strange customs in other groups, such as graves (sometimes very deep) inside caves, as if they were the gateway to a new world or a new life. Or perhaps a return to Mother Earth, who made the grass and the trees grow and from whose womb the precious water of the mountain streams flowed.

His reminiscing is interrupted by a pat on the shoulder indicating that the burial is over and it's time to go home.

He takes one last look at the rocky wall and at the fresh mound at its foot. Then he turns around and walks away with the others.

Traces of Cannibalism

The tiny "cemetery" at La Fasserie described above in a flashback from the past seems to indicate (as do other finds) that the Neanderthals did not merely bury their dead in order to keep them from being ravaged by carnivores; the finds suggest real "funerals," ritual ceremonies.

Perhaps one day we will gain further insight into this aspect of the Neanderthal culture—the sign of a growing sphere of magic in the world of prehistoric man.

Some people feel that it is in this light that certain mysterious finds that might otherwise be mistaken for signs of cannibalism must be interpreted.

Was cannibalism practiced in the past? Two finds, which will be discussed shortly, would suggest that was the case. One is in Krapina, Yugoslavia, and the other in Hortus, France.

Before considering the issue, it should be emphasized that instances of cannibalism have been greatly exaggerated and distorted by biases. Cartoonists who depict explorers in huge caldrons do nothing but play on the widespread belief that primitive people ate human meat. As anthropologist William Arens says in his book *Man-Eater Myth*, in the past the Jews, the Christians, the Irish, the English, the French, and the Germans have accused one another and the peoples colonized and "civilized" by them (Africans, Papuas, Indios, Philippines, Aborigines, etc.) of cannibalism. It is a common accusation used to discredit the adversary.

Cannibalism has existed and does exist, but it has very specific characteristics. Three types can be distinguished:

1) *Nutritional cannibalism.* This occurs when human meat is eaten as food. There are very few reports of this kind of cannibalism. One exceptional case occurred after an airplane crash in the Andes in which, while waiting for rescue, the survivors ate pieces of meat cut (in fine strips by a medical student) from the dead passengers.

It is likely that similar cases, brought on by emergency situations in which food was lacking, have occurred in the past (and may occur in the future). But no human groups or tribes are known to practice cannibalism in the sense that is commonly attributed to the word. It has to be admitted, however, that human flesh roasting over a fire has the same smell as a rack of barbecuing steaks (anyone having taken part in a cremation in an Oriental country knows this). But those who have tasted human meat claim that it is tasteless. The cultural taboo, however, is generally too strong to be broken.

2) *Ritualistic cannibalism.* This is very widespread among many peoples (and may have been even more so in the past). Here the body is not "eaten": certain qualities of the dead person, such as strength or courage, are "taken over"; the dead person is "taken into" the bodies of the living, in order to let him live on or to keep the ghost of the deceased from returning among the living (by eating the brain or the feet). The qualities of a warrior killed in battle are introduced into a newborn child.

Cannibalism in this case consists of eating symbolic portions, perhaps mixed with other substances. In Amazonia, for example, certain tribes pulverize the bones of their dead relatives, mix the powder into a paste with other ingredients, and then give everyone a taste.

It was reported in the news not long ago that an Italian prison inmate was killed by another who then took a bite of his liver.

3) *Pseudo-cannibalism.* Many signs of anthropophagy found on human bones (cutting, scalping, deboning) are actually signs of a second burial. Some people in the past (and today as well) exhume the dead some time after burial in order to free the bones from the dried flesh and

recompose them for definitive burial. In these cases the marks left on the bones may look like traces of cannibalism.

Is this also true of the hominids that preceded us?

We don't really know, but it seems reasonable to assume that the farther we go back in time, the less the ritualism. Therefore, points (2) and (3) (ritual cannibalism and marks left on bones from secondary burials) would certainly tend to be absent in the remote past, because any form of ritualism requires advanced cultural development.

This said, let's look at the signs of cannibalism that have been found, starting with the two Neanderthal sites.

In a site dating back 50,000 years at Hortus, France, two fragments of femurs and three of humeri were found bearing signs of what M.-A. de Lumley feels are intentional fractures of the still fresh bone. Those fractures (made by blows that typically cause lengthwise splitting) suggest attempts to get at the marrow inside. Furthermore, the bones were found together with leftovers from meals.

At Krapina, also, in a site dating back 70,000 years, around thirty fragments of tibias, femurs, radia, and ulnas were found with similar fractures, in addition to the signs of cutting tools used to debone meat. Many experts are inclined to believe, however, that the phenomena can be explained otherwise: carnivores, the pressure of sediments, or even digging during past excavations.

In a Neanderthal site in Marillac, France, bone fragments were found with signs of cuts and burns. Other remains with evident signs of the flesh having been stripped away were found in Klasies, South Africa. The people living there were already *sapiens sapiens*, but very early ones, dating back almost 100,000 years. Many consider this to be the most ancient site of *sapiens sapiens*, even preceding those of Neanderthals. Many fragments of jaws, skulls, and other bones were found with signs of cuts.

A much more recent site (only 6,000 years old) in Fontbrégoua, France, produced various fractured long bones and five skulls with signs of intentional cuts, apparently made with flint tools (someone has interpreted them as ritual scalps).

The signs suggesting cannibalism in all these finds are the marks similar to those found on the bones of animals that have been killed and eaten.

Whether these are cases of nutritional or ritual cannibalism is hard to say. Some people, like Erik Trinkaus, feel that (at least in some cases) the fractures should be attributed simply to the weight of the sediments on the bones. Others, like Tim White, have started reexamining all the remains discovered to date, paying particular attention to traces of cannibalism which may have gone unnoticed at first glance.

Evidence of Violence

The marks on the bones can indicate not only cannibalism, but also forms of violence.

The Neanderthals were evolved, intelligent people, with a culture that for the first time showed respect for the dead through burial. The old and the handicapped were accepted and assisted. But that does not mean that there was no violence in their society. Our societies also care for the old and the handicapped (albeit insufficiently)—we have health care, schools, pensions, and social assistance. But our newspapers are full of headlines about murder and violence.

We have only a few fossil remains attesting to the violence of the Neanderthals: a "Monteggia" fracture, like that which occurs when someone raises an arm to try to ward off a blow; a rib injury due to a pointed instrument; and the lethal thrust of a lance to the pelvic region. (These three episodes are discussed in greater detail in Appendix 2, which describes traumas and diseases of the hominids that can be detected from study of their bones.)

But it is very difficult to draw conclusions from so little information. All we can do is hypothesize that the Neanderthals resembled us in this respect, too; that is, that there were episodes of violence within their groups despite their high level of socialization.

In any case, only with the need to protect possessions (that is, with the development of the sedentary life and farming typical of the Neolithic culture) did occasions for conflict (and, therefore, probably also for violence) develop.

Tens or Hundreds of Millions of Neanderthals?

How many Neanderthals existed? It's hard to say. All together, perhaps many hundreds of millions. According to our calculations, there were from 9 to 90 million in Italy alone.

How do we come upon that figure?

In a preceding chapter we attempted to calculate the approximate number of hominids that lived on the African savannah: Australopithecines, *Homo habilis, Homo erectus,* and so on. Assuming that the population ranged from 10,000 to 100,000 throughout Africa (this is a conservative estimate), the total would be between 2 and 20 billion hominids.

Can the same kind of calculations be made for the Neanderthals? According to the figures of Edward S. Deevey, about one million individuals populated the world 300,000 years ago. As early as 25,000 years ago, that figure had risen to over 3,300,000.

If these estimates are reliable, some extrapolations can be made.

The Neanderthal period ranges approximately from 35,000 to 85,000 years ago. In that time, the global population on the various continents was between 2 and 3 million people, most of whom lived in Africa. We can, therefore, hypothesize that two-thirds lived in Africa and one-third in the rest of the populated areas (the Middle East, Asia, and Europe). That would come to an average population of 2.5 million people in the world at that time, of whom 1.6 million were in Africa and 800,000 outside of it. Attributing only one-sixth of this population to Italy (we must not forget that it was during an ice age and that Italy had a favorable climate), it can be estimated that the population of Italy at the time was 45,000 inhabitants.

If four generations are considered per century, that population must be multiplied by four, for a total of 180,000 Neanderthals per century in Italy.

The Neanderthals were present in Italy for almost 50,000 years, that is, 500 centuries. Multiplying 500 centuries by 180,000 inhabitants yields a total population of 90 million Neanderthals!

Even if this figure were divided by ten (calculating a constant overall population in Italy of only 4,500 inhabitants), there would still have been a population of at least 9 million individuals.

And that is only in Italy. If the number of Neanderthals in all of Europe is calculated, that figure must be multiplied by six. This means there were 50 million to half a billion Neanderthals in Europe. And as many again in the Middle East.

These are, of course, theoretical calculations. The figures may appear excessive, but once analyzed and disaggregated, they are clearly reasonable. The minimum estimate for Italy of 4,500 inhabitants is extremely conservative: equal to the number of inhabitants of a tiny town that is barely on the map.

What makes the figure seem excessive is our lack of familiarity with long periods of time. We consider history in terms of decades or centuries, not in tens of thousands of years. If we then think that the Italian peninsula has been inhabited for millions of years (Monte Poggiolo), this adds another substantial group of ancestors in Italy, although little remains of all these people.

The Neanderthals in Italy

Remains of Neanderthals have been found in only fourteen sites in Italy. Most of these remains are mere fragments, mostly pieces of skulls, jaws, teeth, and bits of other bones.

The sites are scattered throughout the peninsula from Liguria to Apulia (not on the islands). Many of the findings have been made in caves (Caverne delle Fate, Finale Ligure; Buca del Tasso, Camaiore/ Lucca; the caves in Calascio, L'Aquila; Grotta Guattari, Grotta del Fossellone, and Grotta Breuil, Circeo; Grotta Santa Croce, Bari; Grotta Cavallo, Lecce; Grotta del Bambino, Santa Maria di Leuca; Grotta Taddeo and Grotta del Poggio, Marina di Camerota/Salerno).

These sites, or areas around them, may provide other finds in the future. For example, Professor Amilcare Bietti and Dr. Giorgio Manzi of the University of Rome discovered new fragments at the Neanderthal sites at Circeo.

This does not necessarily mean that the Neanderthals lived in caves. It means that the remains in caves are more likely to be preserved. First of all, it is because they are protected from the elements, landslides, and floods; but above all, it is because the ground has not been overturned by plowshares and the roots of trees. The successive layers of deposits act as further protection.

Furthermore, it should not be forgotten that the Neanderthal custom of burying their dead eliminated the principal causes of the destruction of the bones: scavenging animals, sun, rain, trampling, and so forth.

Many other Neanderthal remains have been found in Italy, such as at Fondo Cattie, Lecce; Il Molare shelter, Marina di Camerota/Salerno; Janni di San Calogero, Nicotera/Catanzaro; and San Francesco D'Archi, Reggio Calabria.

Of the millions of Neanderthals who lived in Italy, many may still be discovered. The remains found to date are, however, representative of the various periods. Of course, there are many other sites in Italy in which no bones but other traces of the presence of Neanderthals have been found, particularly tools, e.g., the famous "Mousterian" scrapers.

The most famous Italian find is a skull surrounded by stones, found in the Guattari cave in Circeo in 1939. So much has already been written about this skull that we will not go into the matter. But we would like to point out some aspects that are often overlooked.

The remains were discovered as usual by workers who, upon knocking down a wall while enlarging a basement, found themselves in a cave. Illuminating the interior, they noticed an overturned skull. The site was immediately studied by Alberto Carlo Blanc, the same person who discovered the second skull at Saccopastore.

One very particular characteristic of the cave was that the entrance had collapsed 40,000 years ago. As a result, no one had entered it since then, leaving it perfectly intact. (In the meantime various ice ages had come and gone; modern human beings had appeared; agriculture, the wheel, and bronze had been invented; the world had witnessed the rise and fall of pharoahs, Etruscans, Romans; and the Renaissance and Romanticism had waxed and waned.)

The hominid in Circeo had waited all that time to be discovered in that strange position: the skull with the crushed foramen magnum was overturned and surrounded by stones.

Was this the result of some kind of magic rite? The bones of a leader whose brains had been eaten in some kind of cannibalistic ritual? Or a head left in the center of the rocks and accidentally overturned later? Or more simply, a human skull found by some carnivore and dragged into the cave to be eaten (in that case, the arrangement of the rocks around it would be accidental or previously arranged in that way for some other purpose).

The fact that the Neanderthals buried their dead and probably did not leave them exposed in caves supports the last hypothesis.

And then again, why only the skull and not the skeleton? Could it have been not a burial, but some kind of macabre rite centered around the head of an enemy who had been killed?

Two other human remains—two incomplete mandibles—were found later: one inside the cave itself and one in the sediments outside. The one found inside the cave bears the teeth marks of predators, probably hyenas. Therefore, many now feel that the most credible solution is that the skull was just the leftover of an animal's meal.

The Disappearance of the Neanderthals

The real enigma is the total disappearance of the Neanderthals. Suddenly all traces of the Neanderthals disappear. No Neanderthal remains less than 35,000 years old have been found. Like the dinosaurs, the Neanderthals came to an abrupt and inexplicable end.

From that period onward, the upper strata reveal only *Homo sapiens sapiens*, that is, direct ancestors resembling us in every way. Only a short time later (30,000 years ago), they left sophisticated works of art, such as statuettes and rock paintings. They were practically identical to us.

Those who study the skulls and the characteristics of the ancient *sapiens sapiens* feel that the differences from modern human beings were minor. Raised in a modern family, a child of 30,000 years ago would have substantially the same school and work record as a child born today.

No one can prove such a statement, but it certainly drives home our similarity to the successors of the Neanderthals. No further evolutionary ramifications have taken place since then. The species has remained essentially the same, with only minor diversifications and adaptations.

These new human beings, as we will see in the next chapter, came from Africa. They arrived in Europe at exactly the time—35,000 years ago—when the cold that had gripped the European continent for 20,000 years was giving way to a warmer climate (the so-called interglacial period between Würm II and Würm III).

Perhaps aided on their northward migration by this climatic change, in a relatively brief time the *sapiens sapiens* occupied all the regions formerly inhabited by the Neanderthals and dominated the scene. It was like a cross-fade in the movies: one image faded away as another simultaneously appeared.

Let's see find out more about this cross-fade.

A Few Hypotheses

The only way to approach the enigma of the disappearance of the Neanderthals is by comparing the various hypotheses to proven fact.

The following are the main hypotheses:

1) A direct conflict between the Neanderthals and *sapiens sapiens*, of the kind that occurred a number of times in history when a population with superior technology entered the territory of another population, e.g., the Romans in Gaul, the Spaniards in South America, and the Americans in the West.

2) Disappearance of the Neanderthals by interbreeding, that is, an increasing number of mixed marriages with the *sapiens sapiens* gradually diluted and absorbed the genetic characteristics of the Neanderthals. Genetic mixing is leading to the disappearance of certain ethnic minorities in Europe today.

3) Deadly diseases introduced by the *sapiens sapiens* to which the Neanderthal were not immune. The long isolation because of climate (in certain periods Neanderthals were literally encircled by ice) had not prepared them for contact with biologically different beings. This is what happened to various populations upon the arrival of the whites: the Amazonians, the Eskimos, and the American Indians.

4) The withdrawal of the Neanderthals to increasingly peripheral regions upon the arrival of more organized and more efficient populations and their subsquent extinction, resulting from the lack of local resources and/ or the return of the Würm Glacial Stage (third phase). The Neanderthals disappeared in this period. A similar phenomenon of progressive confinement can be witnessed today with almost all nomadic people (the Bushmen of the Kalahari and the Australian Aborigenes).

5) The direct expropriation of resources by the ever increasing numbers of newcomers. This is what happened in the United States with the extermination of the bison on the prairies, which had provided the basis for sustenance of entire Indian populations.

Other hypotheses exist, but they are less convincing. For example, some claim that the Neanderthals were already on the road to extinction when *sapiens sapiens* came on the scene. Another interesting hypothesis, which cannot be proven, however, is that the interbreeding between the Neanderthals and the *sapiens sapiens* created sterile offspring (as happens today between similar but different species, for example, donkeys and horses). In this way, the Neanderthals were left without descendants while the *sapiens sapiens* were continually reinforced by new arrivals.

Actually, it seems reasonable to assume that there is no single explanation for the disappearance of the Neanderthals, but a set of causes, depending on the place and the time. The various hypotheses listed above may well have combined under certain circumstances, causing local rather than general effects.

But let's see what the artifacts have to say to confirm or disprove these hypotheses.

The Last Spurt

First of all, the migrations of *sapiens sapiens* almost certainly came from the East. Originating in Africa, these populations gradually expanded into eastern Europe before reaching western Europe.

The most ancient artifacts—fragments of modern bones and tools that are quite different from the "Mousterians"—dating back 43,000 years, were found in Bacho Kiro, Bulgaria. These are no longer the traditional round, flat chips, but long chips with parallel edges called blades. The technique is called "Aurignacian," and requires much more skill, but allows for five times as much cutting surface per pound of stone.

Over a period of only a few millennia, finds of this kind shift westward toward present-day Germany, France, and Italy. Unfortunately, doubts about the dating, the brevity of the period, and the scarcity

56. An encounter between Neanderthals and *sapiens sapiens*. Coming from the east (originally from Africa), these new populations quickly supplanted the Neanderthals in Europe. Was this due to competition over resources, conflict, or disease? The rapid disappearance of the Neanderthals is still a mystery.

of finds make it difficult to follow the expansion of *sapiens sapiens* throughout Europe. But all evidence seems to indicate that this progressive penetration originated from the East.

How did the Neanderthals react to this advance?

Of course, we cannot know for sure, but the artifacts found do tell us something: the Neanderthals disappeared only after a few thousand years—not overnight.

This led to the formation of "cultural islands," so to speak, of individuals halfway between the Neanderthal and the *sapiens sapiens*. Artifacts found in various parts of Europe show that the Neanderthals had started to modify their tools. The encounter with *sapiens sapiens*, the realization that their tools were more efficient, and perhaps the competition drove the Neanderthals to seek more modern technology. In France, the "innovative" results are called Chatelperronian; in Italy, Uluzzian; in Western Europe, Szeletian. All are different techniques, but they have one characteristic in common: they are technological responses to the challenges of *sapiens sapiens*.

Not only have more sophisticated "knives" been found, there have also been discoveries of decorated bone tools and ornaments (perforated teeth and shells), which had never been used by the Neanderthals. Many of these artifacts were found together with Neanderthal fossil remains (as at Saint-Cesaire and Arcy-sur-Cure in France and in the Grotta del Cavallo in Italy).

In some sites, the last Neanderthal remains alternated with those of the new *sapiens sapiens*, with tools in successive layers (for example at Roc de Combe in France and at Cueva del Pendo in Spain).

The farther one moves west, the shorter the period of cohabitation. It is as though the migratory waves were slower and weaker at the beginning and became increasingly overwhelming with time, suffocating the Neanderthals more and more rapidly. The Neanderthals seem to have survived in increasingly restricted pockets which were progressively eliminated by the expansion of *sapiens sapiens*.

Various kinds of encounters probably took place between the two peoples (direct conflicts, interbreeding, spread of diseases, forced withdrawals, expropriation of resources, etc.), including cultural contacts, as the new technologies applied to the Neanderthals' tools demonstrate.

This may have occurred in an indirect way, as the result of chain-like transmission within the groups of Neanderthals themselves, similar to what happened in some places with plastics, which spread to primitive populations before the arrival of the whites.

But the Neanderthals' attempts to modernize their technology were insufficient. Such an enormous leap forward was beyond their adaptive capabilities. After more than 50,000 years of isolation and immobility, contact with *sapiens sapiens* proved to be destructive.

The last remains of the Neanderthals date back 30,000 years. There are no bones, no tools, no objects after that. They simply disappeared. Nor is there evidence that fertile crossbreeding with *sapiens sapiens* took place (except for a few enigmatic finds, such as at Hahnöfersand near Hamburg).

Skulls less than 30,000 years old all have the characteristics of modern human beings. Missing are the Neanderthalian features that distinguished the ancient peoples who inhabited Europe for tens of thousands of years.

But who were the *sapiens sapiens* and where did they come from?

14

The Star: *Sapiens sapiens*

The Uppermost Branches

Reconstruction of the latest part of human history takes us back to Africa, which once again offers the most reliable information.

We left Africa practically at the time of the advent of *erectus*, approximately 1.5 million years ago. A new hominid had emerged, the most complete example being KNMWT 15,000, the tall twelve-year-old that had fossilized in the mud of Lake Turkana.

Our discussion then turned to the migrations of *erectus* from the African continent to Asia and Europe, populating vast areas and giving rise to the Neanderthals, among others.

But what happened in Africa in the meantime?

Of course, many things took place in one and a half million years. First of all, the peoples we spoke about at the beginning of the book —the *Australopithecus robustus*, *Australopithecus boisei*, and *Homo habilis*— became extinct.

After generations walked the savannahs of eastern and southern Africa for hundreds of thousands of years, these beings died out. The boy of Lake Turkana had probably met some of them: the *boisei*, which lived in the same area, and perhaps even some *habilis*, which belonged to other branches of the shrub.

As of about 1.2 million years ago, though, *erectus* was the only hominid left in Africa (even if the existence of "evolutionary pockets" of other hominids cannot be ruled out).

Available evidence indicates that in Africa, as elsewhere, *erectus* preserved their characteristics for a long time. Fossil remains dating from a period of over one million years are surprisingly constant in shape.

Yet, the tools became more refined and the first signs of domestication of fire appeared (there are parallels in Asia and Europe). The *erectus* were gradually changing. Different, more evolved characteristics started to appear. The brain expanded. Remains reveal (although in different ways in different areas) an increase in cranial capacity.

What were the African branches of the shrub like at that time?

From Branch to Branch

Not much information is available, but looking at the three or four main finds and moving from one branch of the evolutionary shrub to another, we can see the signs of development. For example, at Olduvai, an individual appeared between 1.1 and 1.2 million years ago who was the first to have a brain volume of over 1000 cc: 1067 cc to be exact. A record.

The features of this individual (called OH9) were still primitive —a large protruding brow, but thinner bones than those of preceding individuals.

Might these have been the kind of *erectus* that left traces of their passage in Kenya at Olorgesailie between 900,000 and 700,000 years ago? In this very important site, 10,000 bifacial tools were found literally covering an area the size of a tennis court, like a pavement made of stone tools.

No fossil bones were found, however. Rather like those at Isernia (it dates almost to the same time), these beings left behind only their works. And an enigma to be solved.

Among other things, the remains of about sixty giant baboons were found at this site. They were terrible beasts, but the Olorgesailie man evidently knew how to deal with them skillfully.

Moving to a higher branch of the African *erectus* shrub—200,000 to 500,000 years ago (the date is uncertain)—brings us to the find in Bodo, Ethiopia.

Unfortunately, the brain volume is unknown because a large part of the skull cap is missing, but many features seem to indicate a transition toward decisively "modern" characteristics destined to emerge later.

These *erectus* were definitely more evolved than their predecessors; some people consider them primitive *sapiens*.

The size of the face is enormous compared to many other finds (in this regard it must be mentioned that the photographs of skulls that appear in books and journals do not give a good idea of the size of the various hominids. Only by seeing the remains or the molds first hand, can one notice the differences, sometime quite striking).

But this find at Bodo has a very special characteristic: the skull is furrowed by a number of tiny cuts. Tim White, who studied them in 1986, feels that they are signs of intentional scalping, carried out with stone tools.

The reason for this is still a mystery: scalping, cannibalism, a ritual or secondary burial are all plausible hypotheses, but there is no proof to confirm any of them.

Two Parallel Lines

Moving to another branch that is even higher up on the African shrub, we come to a famous find: the one at Broken Hill. This hominid is now also considered an early *sapiens*, even though the skull still has primitive features (e.g., one of the most prominent brow ridges ever found). Its cranial capacity is 1280 cc, however, not much smaller than that of modern human beings.

The date of this specimen from Broken Hill (also called Rhodesia Man), in present-day Zambia, where it was found in 1921 during excavations for a mine, is uncertain, but it is estimated to be 130,000 years old.

Looking at this beautiful skull, it is difficult to resist the temptation to compare it with another famous one found in Europe: the one at Saccopastore. Both are of *erectus*, which had evolved into early *sapiens*, and they probably are the same age (130,000 years) and had the same brain volume (1200 and 1300 cc). It is as though different but analogous pressure in Africa and in Europe led to the same model— a model apparently ready to make the last transition toward *sapiens sapiens*.

It is also interesting to note that these two parallel branches developed simultaneously in Europe and in Africa, almost as though in response to a need to select individuals with larger brains, able to improve their survival through greater intellectual, organizational, and oral communication abilities.

As we have seen, this pressure gave way to the Neanderthals in Europe. But the Neanderthals turned out to be a dead branch or a "losing lane" in terms of human evolution. The "winning lane" had long been taken by one of those many African groups that possessed a mixture of primitive and modern characteristics or, as the experts call them, "mosaic forms." More than 100,000 years ago, perhaps even 130,000, primitive forms of *sapiens sapiens*, that is, individuals belonging to the line of modern human beings, were already walking the savannah.

From which part of the shrub did these ancient *sapiens sapiens* come? It is difficult to say. Not enough information is available for the precise reconstruction of their evolutionary course. But it is not difficult to imagine that they descended from forms of archaic *sapiens* that had long appeared in Africa (of which the Broken Hill specimen is probably a dead branch, despite its 1280 cc of cranial volume).

Although some primitive features persist, the remains found of these early *sapiens sapiens* have unmistakably modern characteristics: the shape of the skull, the face, and the volume of the brain. Cranial capacity was already over 1300 cc and soon increased to 1500 cc, that is, the volume of the human brain today.

The Most Ancient Sites

Fossil remains of these surprising "outsiders" have been found not in one site but in many, in eastern and southern Africa. These are the very regions in which Australopithecines, *Homo habilis*, and *Homo erectus* gradually appeared. Once again, Africa was the site of an important event in human history, this time the most important.

These are the localities and the approximate dates of the most ancient sites of *sapiens sapiens*: in eastern Africa: Mumba (80,000–125,000 years old); Kanjera (around 100,000 years old); Singa (uncertain date, perhaps also 100,000 years old); and Omo I (approximately 130,000 years old). In southern Africa: Border Cave (90,000–115,000 years old); and Klasies Rivers Mouth (90,000–125,000 years old). It is astonishing to discover that *sapiens sapiens* lived in such remote times. Only a few years ago, an idea of this kind would have been dismissed. Of course, there is always the possibility of an error, but can all the dates be wrong?

The sites are numerous and contain a number of individuals. Furthermore, the studies have been carried out by various groups of researchers, who all agree on similar dates. The margin of error is, therefore, reasonably small.

Professor Günther Brauer of the University of Hamburg, who studied these remains at length, feels that, although found at great distances from one another, these individuals were probably linked to a common stock. Migration was a consolidated custom, especially when a species was able to move self-sufficiently in the environment.

Thus, it seems that these regions and these individuals triggered the process that led in a relatively brief time to increasingly evolved beings capable of sculpting statues; painting rock walls; and inventing the bow, the wheel, and radar.

At what point did they start to leave Africa on the great adventure that was to lead them to settle throughout the world?

The Qafzeh Case

One very important find was made in the Middle East. Sixteen individuals with evident characteristics of *sapiens sapiens* (which have been defined as pro-Cro-Magnonoid, that is, precursors of European Cro-Magnon) were found in a cave in Qafzeh, Israel.

At first it was thought that this site was 45,000–50,000 years old, but recently this date has been set back to about 92,000 years ago, making these the most ancient human graves discovered so far (much more ancient that those of the Neanderthals). The method used, called thermoluminescence, is very precise (it is explained in Appendix 1). It was applied to tools found in various layers: those found together with skeletons turned out to be 92,000 years old.

The interesting thing is that many individuals were buried in this cave in Qafzeh. One was actually found holding deer antlers. Some scholars consider this a sign of culturally advanced symbolism.

The skulls from Qafzeh clearly show that these people no longer had anything to do with *erectus* (nor with the Neanderthals). They had a high forehead, a small face, and a prominent chin. There were no signs of the telltale visor over the eyes. And the brain volume was approximately 1500 cc. Some primitive features were still present, but the general physiognomy resembled ours.

There were probably innumerable divergences among the *sapiens sapiens* as well. The skulls found in the various sites mentioned (in eastern Africa, southern Africa, and the Middle East) all show different mosaic-like blends of primitive and modern features.

For example, at another site in Israel, at Skhul on Mount Carmel, about ten skeletons of *sapiens sapiens* were found dating back 80,000–100,000 years (in the past this site was also thought to be less ancient). This group seems to be slightly more primitive than that of Qafzeh.

Another individual with even more primitive characteristics—the Man of Galilee—was found in Mugharet el Zuttiyeh. Here the dating seems to be more recent (40,000–70,000 years), though, and this does not help to clarify things.

As so often happens, it is not easy to fit the puzzle together and reconstruct the evolutionary routes with only a few pieces and imprecise dates to go on. It is hoped that more will be known in the future.

In any case, it was this variety which slowly gave rise to the model most suited to facing the great challenge of colonization of the planet.

The Enigma of the Migrations

One question immediately comes to mind: Why did *sapiens sapiens* not migrate more rapidly to Europe and Asia? If they were already present in the Middle East over 90,000 years ago, why did they wait such a long time (approximately 50,000 years) before pushing on to the north and the east?

The first finds of *sapiens sapiens* in Europe and Asia date back only 35,000–40,000 years. It may, of course, be assumed that the most ancient sites have not yet been found. But it is nevertheless odd that all the earliest finds of *sapiens sapiens* in Europe and Asia are from the same period. They seem to reflect a massive wave of migrations which may have taken place during a particularly favorable climatic period (in fact, they coincide with an interglacial period). Then again, migrations were not an automatic event in prehistoric times, but were probably dictated by natural laws: humans follow animals, animals seek out vegetation, and the vegetation changes with the climate.

It may also be that only small groups of *sapiens sapiens* came sporadically to the region in which the site of Qafzeh is located (where they probably started to encounter the Neanderthals coming from Europe. Some very interesting studies on these particular Neanderthals have been carried out by Bernard Vandermeersch and Silvana Condemi).

An interesting fact must also be taken into consideration: the tools of these ancient *sapiens sapiens* were of the "Mousterian" type, that is, similar to those of the Neanderthals. Perhaps the cultural and technological revolution that was to allow these beings to develop their pre-adaptations and assert themselves throughout the planet had not yet taken place.

Why a group of people on the threshold of Europe 92,000 years ago would wait 50,000 years before moving onward is still an open question. But from the moment in which they started to move on, that is, 35,000–40,000 years ago, these *sapiens sapiens* were unrestrainable. Both in Europe and in Asia, ancient populations gradually gave way to the new breed.

It is not clear whether and to what degree interbreeding took place in the various regions. Nor is it clear what kinds of local evolution took place. Some scholars feel that the emergence of *Homo sapiens sapiens* is not the result of a migration originating in Africa which obliterated everything that existed in the rest of the world, but rather the consequence of a series of local evolutions, originating perhaps in more primitive forms (perhaps even *erectus*) which, although affected by interbreeding and migrations, gave rise to independent branches leading toward *sapiens sapiens*.

This is the so-called polycentric theory, based on the finds and sites documenting a number of transitions. This theory, however, is not accepted by the majority of paleontologists who feel that *sapiens sapiens* from Africa superimposed their model through multiple migrations (with possible local influence or successive specialization).

On the other hand, the idea that we all descend from *sapiens sapiens* originating in Africa seems confirmed today by a very original study, carried out by the late Professor Allan Wilson of the University of California at Berkeley in collaboration with his colleagues Rebecca Kann and Mark Stoneking. This study is quite different from anything used in paleontology to date; it studies the placenta of women living today.

The Surprise in the Placenta

In order to comprehend this study, it must be understood that each cell in our body contains at least two kinds of DNA. The first, as is well known, encodes all the genetic information of an individual and is contained in the nucleus.

The other DNA, totally different and independent of the first (and 300,000 times shorter) is found inside the mitochondria, which are tiny organs located within the cell that produce energy for cellular processes.

The origin of these mitochondria is not yet quite clear (some hypothesize that they were initially bacteria associated with the cells). The fact remains that they exist and have existed from very ancient times.

During fertilization, the spermatozoon inserts its DNA into the ovum, but it does not insert its mitochondria. Once the ovum has enveloped the DNA of the spermatozoon, it continues to function thanks to its own mitochondria. Thus, the mitochondria that we have in our cells are those of our mothers, not our fathers.

When we reproduce, the same thing happens. Only the females (and not the males) transmit their mitochondria to their children through the ovum. Thus, there is a female line of DNA inheritance, that of the mitochondria, like a baton that is handed from one female to another.

Of course, if a woman has only male children, the transmission of her mitochondrial DNA is interrupted forever. It's like surnames: in most Western societies, those who have only female children will not see their name passed on to descendants. One generation is all it takes for the line to be broken forever.

So, in what way can mitochondrial DNA help us study our ancestors?

The remarkable thing is that, as a result of the mechanism described above, this DNA does not mix with male DNA. It is transmitted intact (or almost) from one generation to another. And, therefore, it is a marker that can be traced through time.

Studies carried out on this DNA show that it gradually transforms as a result of accidental mutations at the rate of 2–4 percent per million years (there is some debate about these figures). This makes it possible to calculate, albeit in an approximate way, the "distance" between individuals: the greater the similarity between their mitochondrial DNA, the closer their common ancestor.

In this way, many observations can be made about individuals today. Samples of the mitochondrial DNA of the placentas of one hundred and fifty women from various racial groups (in Africa, Europe, Asia, Australia, and New Guinea) were studied after they had given birth. Without going into the technical details, the conclusion is that all the women examined showed a convergence toward the same original line situated in Africa. In other words, the baton is the same, despite the alterations caused over time by periodic mutations (but the maximum diversity was nevertheless only 0.6 percent).

The conclusion is that the mitochondrial DNA in all the women examined converged toward one single woman, who lived in Africa sometime between 150,000 and 300,000 years ago. By extrapolation, we are all descendants of that one woman.

That does not mean that there was a "first" woman (also because this woman obviously had parents and ancestors). Again, the analogy of surnames is fitting: they are like markers indicating a lineage that disappears as soon as there are no male children. Statistically, surnames tend to disappear. The same is true of mitochondria. Going back in time, therefore, one can arrive at a single "mitochondrial surname," as happened in this case.

In conclusion, this study shows that:

1) by going back from mother to mother, we find that all of us converge to one African ancestor;

2) *Homo sapiens sapiens* originated in Africa;

3) it is extremely unlikely that other *Homo sapiens sapiens* deriving from independent and parallel evolutions on other continents exist, as there is no trace of mitochondria of other origin;

4) all the races today diversified from a common strain, which originated in Africa 150,000–300,000 years ago.

As always in research, no result is final. It can always be corrected and improved by new discoveries. But this biochemical study is certainly extremely significant, not least because it confirms data coming from paleontology and has recently been confirmed by further research.

Out to Conquer the World

Other research on the genetics of populations, carried out in particular by Luca Cavalli Sforza of Stanford University, Paolo Menozzi of the University of Parma, and Alberto Piazza of the University of Turin, using various techniques (also involving blood groups), has provided extremely interesting information about the migrations of *sapiens sapiens*.

According to this research, the migration of *sapiens sapiens* from Africa started about 60,000 years ago, dividing into two main streams as shown in Fig. 57. One of these headed southeast, reaching New Guinea and Australia (perhaps 40,000 years ago).

In spite of the lowering of the sea level caused by the ice ages, these islands were still separated by the sea. There were almost 63 miles of water separating Indonesia from Australia.

Did these ancient populations know how to navigate? They must have, since bones and implements dating back 30,000 years have been found in Australia. It is likely that these early human beings used rafts of bamboo (abundant in the area) or hollowed tree trunks for their crafts, as certain populations still do today. But that has yet to be clarified.

The second stream of migration from Africa headed north and divided about 40,000–50,000 years ago into two more streams: one toward Europe (the famous migration which supplanted the Neanderthals between 30,000 and 40,000 years ago) and the other toward the northeast, where it forked again, expanding north toward Siberia and east toward Manchuria and Mongolia.

This last migration reached Korea and Japan, probably also using primitive boats (tools dating back 30,000 years have also been found in Japan).

One curiosity: when did the "almond-shaped" eyes so typical of the Orientals appear? The genetic map of the migrations indicates that the two streams that flowed toward Asia (one toward Mongolia and the other toward Siberia) can still be identified genetically today, despite obvious contacts, interbreeding, and migrations. Thus, it can be hypothesized that a characteristic of this kind had already developed before the separation around 40,000–50,000 years ago.

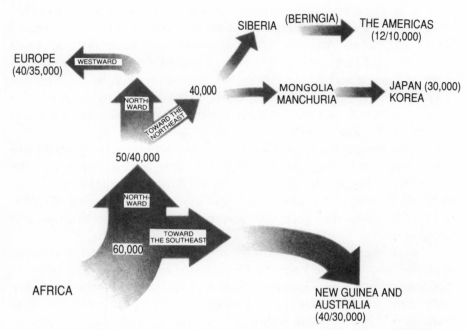

57. According to genetic research on populations, *sapiens sapiens* started to spread from Africa throughout the planet about 60,000 years ago.

The Journey to North America

Starting out from Siberia, the last branch of migrations brought the first human beings to North America (long before the Viking explorers and Christopher Columbus). Taking advantage of favorable climatic conditions, these *sapiens sapiens* managed to cross the Bering Straits thanks to a long tongue of land which had emerged—Beringia.

There is still much uncertainty about the time of these migrations. The most ancient and reliable finds are located in North America and date back approximately 11,500 years (these are the so-called Clovis arrowheads).

Artifacts have also been found in Central and South America, with dates ranging from 17,000 to 32,000 years ago. These remains are, however, very dubious (even though the possibility of such ancient migrations cannot be excluded).

At least two waves of migrations took place. One was around 12,000 or more years ago (the largest and most important, which colonized North and South America). Another minor one took place approximately 10,000 years ago, in which the people settled along the western coast of Canada and Alaska (today the language spoken by these groups is completely different from that of the other Indian tribes, and is, instead, very similar to the language spoken by the Eskimos, who also migrated at the same time).

Thus, leaving Africa perhaps 60,000 years ago, these *sapiens sapiens* were so successful in their migration that they settled throughout the world, from Europe to Japan, Siberia to Australia, Alaska to the tip of South America. In a very short time in evolutionary terms—10,000 to 20,000 years—they genetically overpowered all the previous ancient populations.

These human beings, definitely capable of fishing and navigating, were the first to leave behind extraordinary art works. They were, in fact, exceptional fresco painters. True masterpieces from the time have been found in numerous caves—prehistoric "Sistine Chapels"—with as many as 600 figures, such as those at Lascaux in Dordogne, France.

In order to understand their technique, which has been clarified by modern research on the colors and the instruments used, let's go back in time once more to these caves and watch an ancient painter (who will remain forever anonymous) at work. This is the cave at Lascaux, 17,000 years ago.

A Fantastic Stampede

He runs his hands over the white limestone wall. His fingertips caress and memorize each detail of the rough, slightly concave surface.

His eyes survey the rocky surface. Then he stops and stares at the top of a rocky crest. As the form of a galloping horse takes shape in his mind, the rocky crest transforms itself into one of the hind legs.

Slowly moving the torch made of an evergreen branch, he attentively observes the interplay of light and shadow created by the cave at that point. Gradually the flanks, the head, and the broad belly of the horse take shape in his mind. The figure has been composed. It simply has to be "traced" onto the rock wall.

But the volume of the animal's body cannot be rendered by the natural roughness of the rock. The color, tonality, and shading will be decisive.

By the light of an "oil lamp" made of a concave rock filled with animal fat, the man crushes a few lumps of ocher with a pestle. Slowly the mineral disintegrates into a bright red powder.

With slow, sure gestures, the man mixes the powder with clay and adds some water from a deerskin pouch. He starts to knead the reddish-orange mud.

Suddenly he stops. His face twists into a grimace and a sneeze rings through the dark, silent cave.

He looks up and rubs his nose with the back of his hand, smearing red paint on his face. He is a middle-aged man with a beard and sparkling eyes. He is well known among the groups in the area for his incredible drawing ability and the aura of magic he creates with his paintings. It is a family gift: his father was also considered half-artist, half-witch doctor. He has instructed a couple of "apprentices" with whom he has sketched out other figures in the cave—also horses with enormous bellies and small heads. The frescoes are often made up of a number of figures, each with a precise meaning.

It is curious to note that 90 percent of the animals hunted by these people are reindeer, and yet there is not one on the walls. Horses and giant oxen are the subjects most often painted. It is obvious that they have precise significance (perhaps they are male and female symbols or perhaps they are associated with the ritualism of the hunt, as the frequent arrowheads and darts seem to suggest).

From his crouched position, the man slowly gets to his feet. He is holding a small wooden tray (a kind of pallette) containing the colored mixture. With great care, he climbs onto a wooden scaffolding several feet high, which trembles with his every movement.

The torch tied to the scaffolding wavers, sending light shivering over the walls and the roof of the narrow natural corridor. The horns of oxen, the legs and head of another horse painted beside the scaffolding and curious geometric symbols and dark signs are illuminated for a moment in the darkness.

The painter's expression becomes more intense as he reaches the upper part of the wall where the fresco is to be painted. The moment has come. The man dips his fingers into the reddish mud and starts

58. ". . . With a sure hand, he starts to outline the horse. . . ."

to apply it to the white limestone wall. He uses his thumb and a small bunch of tightly bound grass to spread it.

With determined strokes, the orange patch expands, stretches out, folds back, gains volume, and fades at certain points. The man concentrates on his work. The tension of recreating on the wall what is so clear in his mind is betrayed in the taut lines of his face. The birth of this new figure is reflected in the dilated pupils of his eyes.

At a certain point he holds the oil lamp up to the wall. Every pore has been impregnated with color, still wet and shining.

Without taking his eyes off his work, he bends down to pick up a black stick. It looks like a charcoal pencil, but actually it contains manganese oxide, another opaque black mineral that is often used for frescoes.

With a sure hand, he starts to outline the horse, starting from the leg on the rocky crest. Then the dark line departs from the patch of orange color. An error? No. Using the white background of the wall, the artist gives color and body to the belly and legs, which slowly take shape along with the black hooves. Then he uses another trick to give the idea of depth. The limbs on the far side of the horse are left detached from the body. This gives the impression that the space in between is actually filled by the body itself.

This is a very simple but highly effective trick, which marks the birth of perspective in very remote times. Some figures even seem to be painted in a deliberately distorted way, so as to take into account the observer's point of view.

Now the mane and the snout are highlighted by skillful shading with the black manganese oxide.

The man works with great expertise. No hesitation, no error. Despite the various and uneven surfaces, no corrections are required.

Now the meaning of the two large feather-like shapes previously outlined in ocher becomes clear. They are the spears whizzing past the horse.

The artist adds the finishing touches to the horse's chest. He steps back to look at the final effect.

The work is now complete. The bright wet orange paint shines in the light of the lamp, giving the running horse a magic luminosity. Tomorrow, when the propitiatory ceremony takes place here, the fresco will be dry and all will be able to admire the fluidity and form of this new figure.

The man casts a last glance at his red horse while climbing down from the scaffolding. He squints as if he can almost see it running. A smile spreads over his tired and red-, black-, and orange-spotted face. He blows out the lamp, gathers up his things, and walks toward the entrance of the cave.

He does not feel alone. Around him, in the light of the torch, appear multicolored herds of gigantic oxen, broad-bellied horses, and deer with magnificent antlers. But they disappear as quickly into the silent shadows.

At the entrance, the man rubs the torch to put it out and walks away, throwing a last glance into the cave. The orange horse remains mute and immobile in the dark and its flanks, still shining, will have to await new light to take up its imaginary flight.

For 17,000 years that fresco was forgotten.

Only in the afternoon of September 1940 did some children discover the entrance to the cave hidden by the root of a tree. Lowering themselves into the cave on a rope, they found themselves faced with these ancient paintings shimmering in the dim light. Since then electric light has illuminated them for the world to admire.

Just Like Us

Our story of prehistory could end here, because the people who populated the earth at that time were identical to us. They were like we are today. With the proper education, the child of those *sapiens sapiens* could have been biochemists, writers, or stockbrokers.

Biological evolution, that long road that led from Australopithecines to the people of Lascaux, has come to an end. After over three and a half million years, the footprints at Laetoli, from where we started, have taken us to modern human beings. This is where biological history ends and cultural history begins.

The artifacts found in the sites of *sapiens sapiens* become more abundant and refined. Not only bones and tools have been found, but objects, decorations, and even sculptures: for example, the famous "Venuses"—actually statuettes made out of various materials (clay, ivory, etc.)—schematically representing the female body, often with very Junoesque forms.

59. The extraordinary mammoth ivory sculpture of a *sapiens sapiens* head made 26,000 years ago: the first real three-dimensional portrait of one of our ancestors.

These finds also include an extraordinary head of a *sapiens sapiens* sculpted into the ivory of a mammoth 26,000 years ago. It is the first three-dimensional portrait of one of our ancestors. The face is Nordic and has long straight hair. It could almost be taken for a Russian or a Swede today. The eyes are slanted, very expressive, and highlighted by high cheekbones (Fig. 59).

This man lived in eastern Europe. The statuette was found in a camp close to Dolni Vestonice in the former Czechoslovakia. At first it had been taken for a skillful forgery, but now all analyses confirm its exceptionally ancient origin: 26,000 years.

The interesting thing is that this person, although modern in every respect, still has a strongly pronounced brow. And this is in keeping with the skulls found in that area from that time.

In fact, the *sapiens sapiens* were not all the same. There were probably different kinds and races (some of whom had more pronounced features than others), depending on the place and the time.

The most famous of these is without a doubt the Cro-Magnon, the first to be discovered. Its name (which derives from the French village by the same name close to the site) is often used to indicate all the modern populations of the time. But the *sapiens sapiens* who migrated to Asia, Australia, Siberia, and America were certainly quite different from the French Cro-Magnon (we need only to look at the different kinds of human beings on the earth today to realize what differences there may have been then).

But what kind of life did these people lead?

The sites and the graves of all these *sapiens sapiens* give us a precise idea of their lifestyle.

Prehistoric Renaissance

These were really new people with respect to the Neanderthals. All it takes is a glance at a grave like that at Sungir, northeast of Moscow, dating back 23,000 years, to realize that. The skeleton of the deceased is adorned with bracelets, the teeth of Arctic wolves, and a band of engraved ivory around the forehead. It is wearing trousers, stockings, and boots, the buckles of which have been found. Many other graves contain similar ornaments and headdresses that were not meant merely for protection against the cold. They clearly responded to an esthetic sense and may have been symbols of prestige.

The sites in Russia also bear evidence of huge huts, built with enormous mammoth bones and tusks. The objects found are an indication of the people's new abilities and activities: bone needles for sewing clothes; (decorated) spear throwers for more efficient hunting;

statuettes; new and very sophisticated instruments like arrow stabilizers; ivory buttons; perhaps even musical instruments (a small flute made of bird bones); geometric shapes; dolls; and even jointed statuettes, like the 28,000-year-old figure of a man with movable arms made out of mammoth ivory found in Brno, in the former Czechoslovakia.

A real outburst of cultural activity in only a very short time: just over 10,000 years. A kind of prehistoric Renaissance, which anticipated revolutionary changes in social organization. Some of these objects bear decorations or symbols that have not yet been deciphered, such as geometric shapes, dotted lines, spirals, and anthropomorphous figures with the heads of animals (like the statuette with the lion's head sculpted in Hohlenstein, Germany, 32,000 years ago).

In the absence of sure interpretations of these symbols and signs, diverse hypotheses have been advanced. But caution must be exercised in interpreting objects from so long ago. For example, certain lines engraved on pebbles found in Rochedasse, France, dating back to the Paleolithic period, suggested an ancient calendar. But Professor Francesco D'Errico of the Institute of Human Paleontology in Paris has proven that this cannot be the case. By reproducing the lines on other pebbles and comparing them with the originals under the scanning electron microscope, he has shown that the lines were engraved rapidly one after the other, rather than at regular intervals as one would do for a calendar.

A Prehistoric City

Central-eastern Europe, especially Moravia, is turning out to be one of the most surprising and rich areas for finds of this Paleolithic Renaissance. (Many are far more advanced than finds from the same period in western Europe, perhaps because *sapiens sapiens* had settled there much earlier, building up a minor "civilization.")

There is evidence of this civilization in two sites in the former Czechoslovakia: Predmosti and Dolni Vestonice. Both are between 20,000 and 30,000 years old. Over 40,000 implements and fossil remains of mammoths have been found at Predmosti. This shows that the site was not merely a camp, but probably a rather large settlement.

The remains of at least fifteen (not necessarily contemporaneous) sites have been found at Vestonice on a mountainside overlooking a river. The place has a bottleneck through which all the animals in the area had to pass.

Was this an early "city"? Professor Jan Jelinek of the Anthropos Institute Moravske Muzeum in Brno, who has long been studying these sites, comments:

It is not an exaggeration to speak of a large stable settlement, in the sense that there were many inhabitants and they lived there for a long time. They were certainly not small nomadic groups. The site was strategic because it had control of all the hunters moving north through Moravia along the river to Poland where the source of the raw materials for stone tools was located.

Some of the graves in these sites were very elaborate, others were unadorned. Was this already a sign of social differences?

That we do not know, but there certainly are differences in the graves. A triple grave recently discovered, gives the distinct impression that the individuals did not belong to the same social class.

Such a highly developed culture was already familiar with the technique of baking clay. Did they have pots?

Although these people baked clay, it was used for religious, not household purposes. Only much later, when the society was faced with different circumstances, did the need for pots arise. Thus, there was a pre-adaptation for pottery, in the sense that they knew how to work clay. As in biology, certain pre-adaptations which had initially developed for one reason became useful later for quite another in a different situation. People sometimes speak of "great discoveries," but that is not always exact. There is a difference between the knowledge of a thing and the need which leads to its widespread use. It is interesting to note, for example, that these

60. The 3.7 million years that passed between the footprints at Laetoli and the footprints on the moon are represented here on a twelve-month calendar. The first of January is Laetoli. Only at the end of May does *Homo habilis* come on the scene, chipping the first stones. Socrates, Leonardo da Vinci, and Einstein appear on December 31. We are starting to understand our past, but all the powers of our imagination are not enough to conjure up the future.

Lucy
10 MARCH

1 JANUARY
Laetoli
footprints

28 MAY
Homo habilis

6 JUNE
A. robustus

26 JULY
Homo erectus

30 DECEMBER
Lascaux paintings

21 DECEMBER
Neanderthal Man

31 DECEMBER

6 P.M.
Socrates

10:48 P.M.
Leonardo

11:44 P.M.
Einstein

11:57 P.M.

2000
5000
10000
2000

people already knew how to grind and perforate stone 27,000 years ago, that is, they were aware of a technology that was thought to date from the Neolithic period, almost 20,000 years later, when tools worked in this way were used in agriculture.

The Importance of Pre-Adaptation

This brings us back to pre-adaptation, which allowed for the many trans-formations that have occurred in both biological and cultural evolution.

We came across this phenomenon when we discussed the transition to bipedalism (which took advantage of various kinds of pre-adaptation, signs of which are still visible in many animals today). We also came across it in the many abilities possessed by primates (pre-adaptations to certain social behaviors, to the use of tools, to learning and to language) which were further developed in hominids. Pre-adaptation for the manipulation of fire was seen as essential for the discovery of its use.

Something like pre-adaptation is involved in almost all inventions: they can only be made when there is a cultural pre-adaptation. Similarly, no biological "invention" could ever have occurred suddenly; genetic pre-adaptation was required, whether accidental or designed for some other purpose. At the right moment, it offered a solution to a new problem posed by the environment.

At a certain point in human history, pre-adaptation of cerebral capacities allowed human beings to accelerate their own evolution. Genetic "inventions" (based on reassembly of DNA) were no longer required because they took too long. The new solution called for cultural "inventions" (based on the reassembly of knowledge and ideas). This led to new social, economic, and, finally, political organization.

The Paleolithic period had come to an end. The Mesolithic period (10,000–12,000 years before ours, in which the bow was invented) passed rapidly as did the Neolithic period (which began 8,000 years before our era, with the advent of farming and livestock raising).

During this last period of the Stone Age, human beings started to domesticate animals. The most ancient evidence is a grave found in the Palegawra cave in Iran, dating back 10,000 years, in which numerous canine fossil remains have been found with features resembling those

61. The domestication of dogs may have begun with a wolf cub brought home from the hunt which adapted well to family life. A child holding a pup was found in a burial site dating back to 10,000 B.C.

of the domesticated dog. A grave in Israel from almost the same period contained a boy holding a puppy in his arms.

Domestication may have started when wolf pups were brought home after a hunt and given to the children as playmates. These pups would have grown up considering the tribe as their family. Becoming a part of the tribe, they may later have been bred for their tracking and hunting abilities.

These were the last periods in which stone was used, before the invention of metals, which were technologies destined to revolutionize ancient society.

These last and very interesting periods of prehistory are beyond the scope of this book and would require another volume. Our journey along the path of human evolution has come to an end.

What's Next?

Although we have come to the end of our story, this reconstruction of the past can be altered at any time by the discovery of new data, allowing for new interpretations or weakening old hypotheses.

And that is the fascinating part of research: the constant challenge to go beyond the present state of knowledge. The higher the peaks we climb, the farther we can see. Awareness of this challenge is important in keeping an open mind to new contributions (of which we hope there will be many).

What does the future of human evolution hold? During its latest stages, there has been a remarkable acceleration in cultural and technological development. Inventions have abounded in the most diverse fields. Each has added to the development of human creativity.

It took human beings only a few thousand years to invent the wheel, writing, and mathematics; only a few centuries to invent the printing press, the harpsichord, and film; only a few decades to invent television, computers, and interplanetary probes. What's next?

15

In a Hundred? Thousand? Million Years?

A Look into the Future

Having come to the end of the extraordinary story of the human species, a look at the future is in order, to try to understand what lies ahead.

But this is rather more difficult. Unlike the past, there are no facts to go on—no fossil remains, artifacts, or footprints. The future cannot be reconstructed. But some clues can, nevertheless, help us to formulate hypotheses.

This exercise in imagination will start out with things that are taking place today, and will slowly move on toward things that are farther off and more difficult to imagine.

Let's start by saying that the human species is undergoing important biological changes.

If the paleontologists of the future were to study bones from this period, they would be struck by a sudden change in fossil remains. They would find that individuals are much older (in only a century, the average life expectancy of the Italian population has increased from forty-three years to seventy-two for men and seventy-eight for women) and much taller (in a century, the average height in Italy has increased by five inches). Furthermore, they would find far fewer skeletons of children (infant mortality in Italy a century ago was 400 per thousand in the first five years of life, that is, four out of every ten children died!). Moreover, these paleontologists would come upon an enormous increase in the number of skeletons throughout the world (in a century, the population of our planet has increased from 1.6 billion to over 5 billion inhabitants). They would also notice a striking increase in the incidence of tumors (a typical disease of aging) and of handicaps. (The lifespan of the handicapped has been prolonged by medicine and social organization.)

All this in only one century—a radical change with respect to the past. Any paleontologist would notice it at a glance, simply by looking at some skeletons.

But an even more important change is already underway. For the first time in their long evolution, human beings are altering the ancient laws of nature. They are going against one of the greatest regulators of life: natural selection.

For millions of years, human evolution (and that of all life on earth) was based on this fundamental mechanism. Today, in the advanced industrial societies, it is increasingly easy to break the barrier of selection, as can be seen in the decrease in infant mortality and the increase in the number of handicapped, diseased, and elderly.

From a genetic point of view, this means that the bearers of congenital diseases and "weaknesses" of various sorts can reach reproductive age and circulate their genes through their descendants.

This process is already underway, supported by the social, legal, and moral rules that obviously tend to confirm each individual's right to life and reproduction.

Then again, it would be inconceivable to reestablish the criterion of natural selection. Only a madman would oppose human solidarity toward biologically weaker individuals (though various voluntary forms of euthanasia and therapeutic abortion are developing today). From a genetic point of view, society will tend to become less healthy as a result.

Actually, the mechanism of natural selection does not even hold any longer for brain development. This book has underlined the fundamental importance of the progressive increase in the volume and the complexity of the brain, from 600–800 cc in early *Homo habilis* to 1000–1100 cc in *Homo erectus* and 1300–1500 cc in *Homo sapiens sapiens*.

This selective process, which allowed for the gradual buildup of the brain of modern human beings (thanks to repeated selection for intelligence, language, and imagination), has now come to an end. Insufficient intelligence no longer leads to death. Nor does it seem that the more intelligent individuals reproduce more than the others. Therefore, a further increase in human intelligence cannot be expected through genetic selection.

But an enormous increase in the general development of intelligence, through education and the environment, is predictable. And this brings us to the crux of the matter: technology is replacing natural selection as a factor of adaptation.

Homo technologicus

To take one example: An elderly man today would be able to move faster than a young *habilis* (by using an automobile), lift more weights than ten *erectus* together (by using a crane), and strike down a hundred Neanderthals with one flick of the finger (by using a missile).

In other words, selection based on physical abilities only works in a primitive environment, not in an industrial one. In industrial society, muscular power is needed only for manual labor, which is becoming obsolete. Or for sport—where the law of natural selection still reigns—but which is no more than a game. It is no longer important in our society to be strong, fast, or muscular. It is not even important to have good teeth (we have dentures), good sight (we have spectacles), or good hearing (we have hearing aids).

In some way, technology provides us with a new musculature, skeleton, and nervous system, making it possible to be much more efficient not only in terms of strength and speed, but also in terms of perception (with telescopes and microscopes, things can be seen that were invisible to hominids; with radio, even people with hearing difficulties can hear thousands of miles away).

The same holds for intelligence. While natural selection for genetic, that is, innate, intelligence has ceased, the possibility of developing the mental capacities (not only of individuals but of entire populations) through education has increased so much that results far superior to those of any genetic selection may be achieved. Mass education makes it possible to take advantage of intelligence and brilliance which may not even have come to the fore in the past.

This seems to be the tendency in human evolution. Natural selection (both physical and cerebral) is finished and technological selection and, more generally, cultural selection have taken over. Today, an individual or a society can multiply its strength or intellectual potential through the development and correct use of technology (education is a byproduct of automated production, as it allows young people to remain in school longer).

This shift from physical force to cultural supremacy has also brought about an important change with respect to the past. The old competition for physical reproduction has been increasingly supplanted by competition for cultural reproduction.

In fact, the people who succeed today are not the strongest or those with the most children; they are those who manage to obtain the greatest consensus within the group or the population in which they live. It will not be their genes that are present in future generations, as has always been the case, but the products of their intelligence: their ideas, their discoveries, and their works.

This is exactly the opposite of the evolutionary mechanisms described in this book. Individuals can die without children, but their works and ideas can live on and proliferate, becoming their "children"— their "chromosomes," so to speak. This is what has happened with all the great figures of the past. Their ideas, their works, and their discoveries have profoundly influenced our recent evolution—and in the most diverse fields, from politics to religion, research, economy, art, and so on. Competition continues to exist, but in a very different form from the competition of the ancient past.

But what else could the paleontologists of the future discover by examining our skeletons?

They could, for example, discover increased interbreeding. Mobility within countries and continents will bring about ever greater genetic mixing, even among different races. This has already occurred in certain countries (the United States and Brazil, for example), but will take place increasingly all over the world.

In Europe, the growing number of immigrants from the Arab countries, from Africa, and from Asia still form a "foreign body" from a genetic point of view. It is likely that barriers and opposition to racial mixing will become stronger in the short term, but in the long term (centuries and millennia) it will probably be an inevitable process. Asian and African characteristics will prevail, dominating the scene numerically as well. But cultural efforts to preserve ethnic "diversity" (local dialects and traditions, for example) are to be expected, just as attempts are being made to save the panda or the grizzly bear from extinction.

But let's get back to technology as a substitute for natural selection. Technology is also replacing natural selection in another important way: it is intervening directly in biology.

In fact, although advanced societies are envisaged with a growing number of old, ill, and handicapped, that does not mean that those societies will be made up of more and more people in wheelchairs and under oxygen tents. Once again, technology could profoundly change that kind of scenario by "correcting" nature. Some techniques that are being developed today could upset the very rules that have governed for millions of years.

The New Biological Technologies

For example, it is now possible, at the level of animal experimentation, to reverse paralyses: the limb of a mouse can be rehabilitated by means of an embryonic transplant. It is hoped that in a not too distant future this revolutionary medical technique (consisting of the transplant of embryonic nerve cells, which are extremely adaptable) may be used to reactivate the transmission of nerve impulses in the spinal marrow of paraplegics, allowing them to walk again. Similar experiments have been carried out (successfully) on the optic nerve.

This is only one example. But as the study of embryology progresses, it will provide greater insight into the repair of injuries through reactivation of tissues, or even the regeneration of organs (as is being done with the liver), or missing limbs.

The future prospects offered by biology are so overwhelming that the absence of natural selection loses importance in comparison. Human beings have finally put their hands on the key to genetics. As a result, it will be increasingly easy for them to intervene and control the randomness of accidental mutations. This change is so different from anything that ever occurred in the past that we cannot even imagine its scope.

Past evolution took an incredibly long time, advancing through accidental mutations, adaptations, and selection. In the future, genetic engineering, that is, the ability to modify DNA sequences, thereby creating a different "response" of the genes than the one prescribed by natural genetic planning, will make artificial evolution possible. This new kind of evolution could lead in unforeseen directions . . . and in a very short time.

As is well known, genetic engineering is starting to flourish. For now, it is still limited to the genetic modification of bacteria and the "patenting" of new organisms that did not exist before (and that may never have been generated by nature). But the first applications to the human species are about to be undertaken.

For some time in the future, these will be limited to simple manipulations related to medical therapy. But the genetic mapping of human beings is progressing rapidly, as are the techniques for cutting and reassembling DNA squences.

This field is complex and complicated, indeed, but the advances being made are enormous. No one can foretell the future, but it is only a matter of time. Whether it be years, decades, or centuries—all are negligible in evolutionary terms.

Many biological manipulations are already applicable (and being applied) today. Others are on the horizon and will pose unprecedented ethical and legal problems.

Suppose, for example, that one actually manages to reactivate the nervous transmission of the spinal marrow by a transplant of embryonic cells. What mother would not wish this therapy to be applied immediately to her child paralyzed in an accident? But this (and other cases like it) would require the sacrifice of an embryo. And this could lead to an embryo market supported by spontaneous or surgical abortion. (Then again, a woman of reproductive age could produce her own embryo.)

If bioethics were to prohibit such biological wizardry, how would those deprived of these therapies react? For example, what would a mother say, who is told by a committee that her child cannot be offered this kind of treatment because it is based on a method considered unethical?

These kinds of problems will probably arise with increasing frequency in the future. And the issue of embryonic transplants is only the first in a series of dilemmas that could arise from the unpredictable developments of medical technologies, in particular, of genetic engineering.

A few examples will suffice to give an idea of the kind of escalation that could result: If genetic engineering were to develop a way to cure diabetes, very few people would probably be opposed—especially if it were limited to modifying the individuals' genetic response, avoiding the transmission of mutations to descendants (in other words, the changes would not be hereditary; they would be buried with the individuals). It is likely that that kind of therapeutic engineering could also be accepted for hemophilia or for other genetic diseases.

And what if a genetic method were found to "vaccinate" individuals against tooth decay, so that their teeth would stay healthy all their lives? Or what if a way were found to improve individuals' genetic make-up to make them more resistant to certain diseases (including tumors) by, for example, strengthening their immune systems? Or slowing down the aging process? Or what if a way were found to enhance the biochemistry of memory (or of other mental faculties) in order to counterbalance the deterioration caused by aging?

This would (technology permitting) gradually lead to the progressive manipulation of the individual. Holding the key to genetic development means that human beings will soon be able to do far more than natural selection was able to do for millennia.

We are not about to go into this field, which is full of ethical and emotional issues. But it is important to underline that for the first time in human history, techniques are being developed that can modify (or even model) human characteristics, implying some kind of artificial selection.

What is striking is that these techniques of artificial selection are being developed at exactly the time in human evolution when natural selection is disappearing, almost as if nature were "handing over the baton" to technology.

But will some kind of artificial selection be able to develop in the future, despite the foreseeable technical and ethical difficulties?

Is it possible that, without giving in to the temptation to decide which genetic programs are "right" (since the strength of nature has always been its immense variety of genetic combinations—even the apparently most useless or "wrong" ones sometimes turned out to be a prelude to new and more successful adaptations), human beings will correct at least some of the consequences of the interruption of natural selection which they themselves have caused?

The Fleeting Dimension of Time

When speaking of what may take place in the future, it is important to clarify what is meant by the word "future"—whether the next years, decades, centuries, or millennia.

We are actually quite unfamiliar with the parameter of time because our perception is inevitably linked to the direct (or cultural) experience of the world in which we live. When we speak of "future," we are usually not looking too far ahead: to the year 2000 or 3000 at the most.

But what will the earth be like in the year 8740? Or 27,832? Or 58,521?

As we have seen in this book, 10,000 years are the blink of an eye in terms of human evolution. The Neanderthal period (which was very short) lasted 50,000 to 60,000 years. The *erectus* lived twenty times that long: over one million years. What will human beings be like in the year 148,782? Or 636,418? Or 1,134,420?

If we put ourselves into that kind of perspective, all reasoning becomes vain and our current technological predictions ridiculous. We will never be able to imagine what transformations the planet, much less the human race, could undergo (given, of course, that humanity does not annihilate itself and its entire ecosystem in the meantime through nuclear conflict, poisoning, pollution, ozone holes, or climatic crises. These are all possible—although not as probable as many fear).

Moreover, when we look at the future, no matter how technological it seems to be, with space stations and computers, extraordinary medicines and robots, human beings are always in the center—human beings that resemble us, wearing silver suits, perhaps, or with mini-jets on their backs, but resembling us nevertheless.

In other words, we have difficulty imagining people of the future who are dissimilar or even totally different from the way we are today. We accept the fact that over millions of years evolution has gradually led from ape-like beings to modern human beings. But why should we continue to change? Have we not reached perfection?

Do the statues of Praxiteles and the frescoes of Michelangelo not represent models of human beings to be preserved forever? What madman would want to change those bodies or those faces?

After all, human genius reached its peak with people like Leonardo da Vinci, Bach, and Einstein, didn't it? Our instinctive reaction is to think so. But in only 50,000 years some of our descendants will be as different from these ideals as we are from the Neanderthals. Perhaps even more different, as the changes in the future will not be as linear as those of the past. In all fields, they will be subject to the pressure of increasing acceleration.

It has been calculated that the volume of knowledge currently doubles every 10 to 20 years. If that is the case, acceleration could be such that in terms of knowledge, the future decades will be equal to centuries, millennia equal to millions or billions of years. So, 50,000 years could be the equivalent (from the point of view of change) of 50,000 million or 50,000 billion years.

The way in which human beings will be transformed by such a technological maelstrom is difficult to say. It is even more difficult to predict what their lives will be like. These time frames are beyond the scope of human imagination.

Even the most vivid fantasies are, on the whole, no more than banalities. For example, we can imagine that space will become (and it certainly will) an important new resource, a new habitable dimension thanks to large space stations (capable of hosting millions of people) or the colonization of the solar system, carried out by breaking up planets and reassembling them in suitable ways (a process that some physicists are already hypothesizing today).

We can also imagine that we will travel through the galaxy one day and that individuals will change biologically in order to live in space (perhaps losing their capacity to live on the earth because of the loss of muscle and bone brought on by the decreased gravity).

We may imagine that intelligent androids (humanized robots with their own likes and dislikes and complex reactions) will be built in the future, if software is oriented toward the achievement not only of artificial intelligence, but also of artificial emotions—something that is already conceivable today. We can even imagine encounters with extraterrestrial civilizations.

But such fantasies will always fall far short of reality. Why can we be so sure about that? The great nineteenth-century science fiction writer, Jules Verne, could, with his brilliant imagination, anticipate submarines, a trip to the moon, or a descent to the center of the earth. But he could not imagine things—the discovery of which was not even very remote—such as radio waves, X-rays, relativity, and quantum mechanics, not to mention nuclear energy, electronics, and genetic engineering.

The many and diverse "surprises" that this last century has offered us clearly demonstrate that just when we think we know everything, we discover that we actually know nothing. The thought of the discoveries, revolutions, and inventions that lie ahead boggles the mind. But where will this dizzying technological evolution lead? Toward what kind of world and what kind of human beings?

The Spread of Intelligence

No one can answer these questions. There are too many variables and unknowns to attempt any kind of prediction. There is, however, one constant throughout the story which may trigger our imaginations.

One of the things that can be observed throughout the story of human development is the gradual increase in intelligence: from bacteria, to fish, to primates, this process has been relentless. The end result, the human brain, is the most complex and sophisticated object known today.

The human brain, composed of the atoms emitted during the explosion of the supernovas, is basically a piece of universe that can look back on its history, ponder itself, and start to direct its own future.

This nucleus of intelligence has continued to develop by creating networks and supernetworks with other brains, that is, through culture. And now it is starting to spread to matter, for example, by putting together molecules of silicon and other minerals to build intelligent machines. Some of these intelligent machines (although still very primitive) are already leaving the solar system to travel through interplanetary space.

Furthermore, this brain is learning to "assemble" other molecules—those that make up the genetic code—in order to modify and (perhaps) construct new forms of intelligence: biological, bio-electronic, or other forms as yet unknown.

In other words, a very slow process that took 15 to 20 billion years has produced intelligent organisms that are now able to transfer new kinds of intelligence to matter rapidly, in a continuing and unending spiral. This fantastic ability to spread to other matter may accelerate in the next hundred, thousand, million, or billion years.

Where will it lead?

A science fiction writer might imagine an "all intelligent" universe, a thinking superentity, like a huge superbrain. But let's stop there. These fantasies could lead us off course. Certainly that kind of prospect would bring some sense into the long and slow evolution of intelligence. But it would also have to be reconciled with the Second Principle of Thermodynamics, which calls for the progressive degradation (sigh!) of the energy potential of the universe and, therefore, calls for the death of all systems that consume energy, including intelligence. But then again, our present knowledge may be updated in this field, too.

In any case, our lives are but thousandths of seconds in the time frame of evolution, both past and future. We can only conceive of our future in terms of those few years or decades that are the measure of our existence. A very limited future, indeed, that allows us, nevertheless, to observe for a short time this extraordinary game of the assembly of molecules and knowledge under way on our planet.

An explosion of knowledge in all fields, including that of paleoanthropological research, is to be expected. And we would certainly like to be around the day a better understanding of the story we have tried to tell will be possible.

Appendices

Appendix 1

How to Date Fossils

A Fossil Is Like a Newspaper

There are at least twenty methods for dating fossils, based on various principles and involving chemistry, magnetics, particle physics, paleontology, and geology. Despite the diversity, all methods can be divided into two main categories: those for *absolute* dating and those for *relative* dating.

An analogy may suffice to illustrate the difference. If we were to find a newspaper in the attic, we would be able to ascertain its date in two ways: by looking directly at the date at the top of the page and by reading the news, trying to assess when the events took place.

Fossil dating methods work in the same way. Some give precise dates (called *absolute* dates), while others suggest the age of a fossil by linking it to other fossils or strata or special compounds in the sediments (*relative* dates).

The Pages of the Past

Unfortunately, fossils are not found with a tag bearing the date. Paleontologists have to base their estimates on external data, much as investigators do after a crime. The first important clue is the site.

290 The Extraordinary Story of Human Origins

For example, if two fragments of bone are found—one close to the surface and the other several feet underground—it is logical to assume that the latter was buried under sediments at a more ancient time than the former was (in that sediments normally accumulate in progressive layers). If a person puts the daily newspaper aside onto a chair after reading it, at the end of a month there will be a stack with the older newspapers at the bottom and the more recent ones at the top. The same thing occurs with geological strata; the lower layers (and the bones they contain) are generally older.

Of course, there are exceptions. In areas in which there has been geological upheaval, faulting or shifting of sediments, dating is complicated indeed, but the principle is nevertheless widely relied upon in paleontology.

The division of the Paleolithic era into "Lower," "Middle," and "Upper" periods follows this criterion: the "Lower" Paleolithic period is the most ancient (and is found in the lower layers), and so on.

With one exception, i.e., varves, this method only allows for approximate dating. Varves are those fine strata of seasonally alternate light and dark sediments left by receding glaciers. The objects and bones found in varves can be dated using a method similar to that of counting the rings of trees, but it is applicable only for a short period of time (from 7,000 to 27,000 years) and in certain deposits.

In most cases, excavation is carried out in places where there have never been glaciers. How does one go about dating then?

From Prehistoric Horses to Horse Power

A good method is to start with the fossil itself. Animals and plants evolve with time, changing their structure and appearance. Thus, the anatomy of the fossil can often reveal its age.

To give an example, we could establish dates in our century by looking at the exterior (and the mechanics) of automobiles, starting with the first Model Ts and going up to today's Maseratis and Ferraris. The same is often done in paleonology: the body structure of animals changed to adapt to their surroundings and these changes can be used to establish dates.

A famous example is the horse. About 50 million years ago it was about as big as a fox (*Hyracotherium*); its front hooves were divided into four sectors and the back hooves into three. It gradually evolved through forms with three sectors to the single structure and the other features of the modern horse, which appeared about two million years ago. The 50 million years of the history of the horse have been well documented by fossils. In cases like this, the shape of a fossil itself gives an idea of its age.

Some other animals which evolved three to four times as fast as the horse provide even more precise points of reference for dating: for example, rodents are considered "prehistoric calendars" from the moment they appeared approximately 500,000 years ago. Their teeth, particularly the form of their molars, precisely indicate the period in which they lived.

Fossils of these rapidly evolving animals—very common and so precious to the paleontologist—are called "index fossils." Other index fossils are the famous ammonites of the Jurassic period (in which the brontosaurus lived) and various kinds of trilobites of the more ancient Paleozoic periods.

Another very useful tracer for dating strata (and, therefore, the fossils they contain) is pollen. Fossil pollen providing information on the kind of plants that lived at a given time are often found. Not only can dates be established, as in the case of animal fossils, but paleoenvironments (that is, prehistoric environments) and paleoclimates can be precisely reconstructed. Sometimes even the habits of prehistoric man can be deduced. For example, in a Neanderthal grave in Iraq, pollen analysis revealed that flowers had been placed around the deceased.

Palynology (i.e., the branch of science that studies pollens) brings to mind some techniques used by the police scientific squad in detective movies: just a few minute grains of pollen can provide an enormous amount of information.

The investigator often has more practical evidence to go on. Just as the time at which a victim's watch was smashed helps to unmask the murderer in mystery stories, so do "clocks" in paleontology provide essential details.

A well-known example are the rings of trees mentioned earlier. Whenever tree trunks are found intact in sediment deposits after thousands of years, the past can be reconstructed by merely lining the trees up as in a game of dominoes. "Index rings" can be found by looking at the characteristics of the individual rings of a tree (which vary from season to season and from year to year in thickness and in color). This method has proven particularly useful in the dating of prehistoric palafitte villages; they are built on piles, which are basically tree trunks. The ancient sequoias in North America make it possible to go back thousands of years.

Fossil Compasses

While the study of the rings of tree trunks (dendrochronology) does not allow us to go back farther than approximately 7,000 years, another "stopped clock" dates much farther back.

We might call it a "stone clock." During rock formation and compression, iron oxides are sometimes trapped, oriented according to the magnetic field of the time. They form tiny compasses. But magnetic north has been inverted a number of times in the course of history. That is to say that in certain periods the needle of the compass would have pointed to what we know as south and not north. Researchers have discovered that all they have to do is look for these natural compasses in certain clay soils, travertine, or volcanic rock to get an indication of time (in the last four million years, the direction of magnetic north has been inverted twice).

Thus, magnetic analysis of rock or sediment can reveal its time of formation and age. This method, called *paleomagnetism*, has provided important information for dating, although there is still some skepticism about the inversion of magnetic polarity in some periods.

Mammoths and Automobiles

How does the famous method involving carbon-14 actually work? An example is very useful here.

Imagine going to a service station, asking the attendant to fill up the tank, and leaving the motor running while the tank is being filled. Theoretically (in the absence of technical difficulties), if gas were continually poured into the tank, the motor could continue to run forever. The meters on the dashboard would always signal a full tank.

Now, if the attendant were to stop filling the tank, the motor would continue to run for some time, but there would be progressively less gas in the tank and the gas meter would drop until the motor would stop running. Now, a bone, like any other organic substance, is simply a natural "tank" for various kinds of atoms, among them a very special isotope of carbon: carbon-14 (written ^{14}C).

Carbon-14 is a variation of ordinary carbon, an atom commonly found in nature and typical of all living organisms (we are made up of hydrogen, oxygen, carbon, nitrogen, calcium, etc.). By breathing, drinking, and eating, living organisms continually absorb ordinary carbon and carbon-14, which is deposited throughout the body and naturally in the bones.

But, as mentioned earlier, carbon-14 is a variation of ordinary carbon: it is radioactive and decays at a constant rate in time. After approximately 5,730 years, half of the original quantity of ^{14}C has disappeared. Thus, each living organism can be imagined to be a car with the motor running. By drinking, breathing, and eating, the organism manages to keep its tank full of ^{14}C. Radioactive decay is not noticed because it is constantly replaced by raw material. But once an animal dies, the stock is no longer replenished and the quantity of carbon-14 accumulated up to that time gradually diminishes—just like the gas in the tank of the car.

When a fossil bone is found, all the researcher has to do is look at the "fuel meter" to find out how much ^{14}C is still in the tank (residual radioactivity), and a simple calculation will give the time at which the animal died.

The carbon-14 technique also works for wood, peat, and shells—basically, any substance containing carbon. For example, it works on fabrics made of plant fibers. In fact, this technique revealed that the fabric of the Holy Shroud of Turin was manufactured during the Middle Ages.

Using this method, only animal and plant remains up to 35,000–40,000 years old can be dated. Further development of the technique, involving the use of a particle accelerator, now makes it possible to date up to 80,000 years back and avoids the partial destruction of the material to be analyzed, which was required in the past.

A Candle from the Past

The carbon-14 method is something like a cigarette butt or a burnt-down candle found by the police during investigations. The length of the candle reveals the length of time it burned.

In paleontology, there are other kinds of candles that can provide clues. One of these is the so-called argon-potassium method. An isotope (a radioactive variant) of potassium tends to turn into a gas (argon) at a constant rate. A very simple analogy is a glass of water that slowly evaporates.

Each human body contains just less than half a kilo of potassium. Of this half a kilo, 20 milligrams are radioactive (that is, potassium 40). Though we are unaware of it, these 20 milligrams turn into argon in our bodies at the rate of 500 atoms per second; the same thing happens in rocks.

This means that if the quantity of argon that has been formed can be measured, the time at which the process began and, thus, the age of the rock can be calculated. It's almost like using an hourglass.

Unfortunately, this method can only be applied to certain kinds of rocks such as lava. In liquid lava or melted rock, argon 40 is freed during cooling and trapped inside the rock. Thus, in the laboratory the rocks just have to be pulverized and the amount of gas measured.

This method is very precise and can be used to go as far back as the earth's formation (around 4.5 billion years ago). Unfortunately, it can only be used on rocks that are at least several hundred thousand or a million years old, depending on the amount of potassium in the rock. Furthermore, it doesn't work on fossils. In order to date fossils, reference must be made to geological strata containing the rock. This results in dating of the following kind: "older than x years" or "younger than x years" or "at least x years," etc. In other words, it would be like discovering the age of a person in a family by establishing the age of the brothers and sisters. This is the system that has been used to date the hominids from Olduvai, Lucy in Ethiopia, and many other finds.

The Flash of Time

In the past, the fossil bones of hominids were often dated in relation to the stone tools found during excavation. The more primitive the tools, the older the find. But this rule does not always hold true: there are still populations in Papua New Guinea who fashion stone hand axes in the same way as our ancestors did a million years ago, but who polish them in a way typical of the Neolithic period. What would the archeologists of the future think if they were to come upon one?

There is a way of establishing the age not only of household objects (such as vases and terracotta containers) but also of flint tools made by prehistoric beings. But one condition has to be fulfilled: they have to have been sufficiently heated, either through firing (as for vases) or through exposure to fire (as for some tools abandoned in a hearth).

This method, called *thermoluminescence,* is based on the fact that the electrons that, with time, are trapped in "holes" in the crystal structures of some minerals (such as quartz) are freed through heating. When a flint scraper is thrown into the fire or a pot fired, these "holes" are emptied, only to be filled by other electrons; the more time passes, the more electrons accumulate. Thus, the number of electrons trapped in the structure tell the researcher how long ago the firing of the vase or the heating of the flint took place.

The technique used is fascinating: the piece is heated once again to 300–500° C; this causes the emission of light (caused by the electrons that are freed). The older the piece, the more intense the flash of light.

The Japanese Touch

Since these dating techniques often entail the partial destruction of the find, they are, of course, not suited for certain unique pieces such as the skull of a hominid.

This dilemma was solved by Professor Yokoyama, who developed a dating technique that does not come into physical contact with the find: gamma ray spectrometery. It is based on the following principle. When fossilization occurs, the original material is replaced, particle by particle, molecule by molecule, by minerals such as calcite or silica contained in the percolation waters that filter through the bone. But water also contains other elements, such as radioactive uranium, which accumulates in tiny quantities in the bone. This produces natural radiation; it is not dangerous. In time, however, uranium turns into thorium, which is also radioactive and both give off what is called gamma radiation. By measuring the amount of gamma radiation emitted, the quantity of uranium that has been transformed into thorium and, thus, the age of the fossil can be calculated.

An hourglass may be a useful example, here, too. The quantity of sand increases in one part as it decreases in the other.

This technique is theoretically very precise and can be used to go back as far as 400,000 years for the dating of volcanic rocks, fossil bones, shells, travertine stone, stalactites, and so on.

The list of dating methods could be lengthened with techniques such as the "racemization" of some amino acids contained in bones (that gradually change their structure with time and therefore make dating possible), the so-called Fission Track method (able to go back 700,000 years), and numerous others. We have chosen to describe only the more common and those most frequently used in paleontology today.

Appendix 2

Trauma and Disease Revealed in Bones

Delayed Diagnosis

Again and again we have emphasized in this book how much data can be gained from fossil bones: they are literally gold mines of information on the age of individuals, the period in which they lived, their evolutionary kinship, nutritional habits, sex, musculature, gait, climbing ability, intelligence, phonation, cannibalistic traits, and so on.

But bones can also reveal the diseases, disabilities, and traumas suffered by our ancient ancestors. It is surprising to discover that they were basically the same ailments that plague us today and that diagnosis can be carried out in the same way.

With regard to fractures, for example, "reduction" of a fracture, that is, realignment of the bone segments, was unknown at the time. Therefore, setting caused overlapping and deformations, but the fracture itself posed serious problems of survival.

At Swartkrans in South Africa, an iliac bone (this is part of the pelvis) of an *Australopithecus robustus* has been found which bears signs of dislocation of the hip. Analyses have shown that the lesion was the result of an accident (perhaps a fall onto the heels). Although the Australopithecine managed to survive, thanks to a shift in the head of the femur, creating another articular cavity, he must have had a painful limp and that would have been a serious handicap for anyone living on the savannah.

Other fossil bones have shown traces of hydrocephalus (water on the brain), knock-knees, osteomyelitis, bone lesions caused by parasitic

62. A fracture must have been a very serious problem in prehistoric times. Only social behavior and group organization allowed an individual deprived of the use of a leg, for instance, to survive.

diseases, and arthrosis (the last, often devastating, was not the result of the humidity of the caves, as is frequently believed, but of the accumulation of microtraumas typical of an active life).

Traces of tumors have been found on femurs, mandibles, and skulls (in the Lazaret cave a piece of skull cap bears typical signs of meningioma, almost certainly the cause of death of the person affected by it: a nine-year-old boy).

Of course, there was a whole range of other ailments which leave no fossil traces because they involve organs or soft tissues: fevers, intoxications, wounds, viral infections, burns, torn muscles, poisonings, pneumonia, gastritis, nephritis, and tumors, as well as damage to the cartilage, the ligaments, and so on.

Traces of Violence

While bones often reveal traces of violence that may have been caused by conflicts between individuals, there is much controversy about the question of aggression among hominids. They—especially Australopithecines—were formerly believed to be very violent and brutal, but that view is changing today. Philip Tobias states:

> Some people think that the Australopithecines were bloodthirsty and aggressive. This idea was initially put forth by my professor, Raymond Dart. I, on the contrary, think that these beings were very gentle and kind. For example, despite their aggressive traits, gorillas are very gentle and I think that hominids were, too.

In other words, it may be wrong to use labels such as "good" or "bad" in referring to these beings. In nature, as in human society, conflict (or the lack of it) depends on the circumstances.

At times there is a tendency to go to the other extreme: the myth of the "savage" who is intrinsically good (and the one about modern human beings who have been made violent by a society of cars and money). Some anthropologists, who have studied nomadic groups of hunter-gatherers in South Africa, have recorded a rather high homicide rate (some actually say that, proportionately, there are more murders among these populations than in New York). It is likely that the hominids, like modern human beings, were neither particularly good nor particularly bad; they probably lived together peacefully most of the time, but had occasional violent conflicts.

We have direct evidence of one of these episodes: the fractured mandible of a Pithecanthropine *Homo erectus* child who lived approximately 500,000 years ago in Java.

The type of fracture rules out indirect causes such as a fall; it seems that the child received a violent blow to the right side of the face. The fractured ends of the mandible overlapped, probably creating considerable mastication problems, but not sufficiently serious to lead to death. When the child died at the approximate age of twelve, the fracture had healed.

Was the fracture caused by a violent parent (and left-handed at that)? Or some other member of the group without too much concern for the child's age? Many believe that this is the most ancient evidence of violence ever found. But the paucity of fossil evidence makes it impossible to prove that it was an act of violence. As in a police investigation, more convincing proof is needed.

Other fossil finds from the Neanderthal period attest to violence. Three examples can be given. The first comes from the Neander Valley and belongs to the very first Neanderthal ever found. The bones of the left arm show typical signs of a "Monteggia fracture": perhaps the arm was raised to ward off a blow. In any case, the result was a high fracture of the ulna and the dislocation of the head of the radius. Although the fracture healed, the person was probably no longer able to raise the left arm above chin level.

Of course, it is possible that the fracture occurred during a fall, but one tiny clue tends to validate the first hypothesis: it seems that the left arm (or side of the body) is injured more frequently since aggressors usually strike with the right arm. T. Dale Stewart noticed this after studying countless fossil finds.

The second example, from Shanidar, leaves no room for doubt: the remains of the Neanderthal reveal traces of a foreign body (perhaps a spear) that penetrated from the left, grazed the ribs, and probably entered the lung. The signs of scar tissue suggest that this injury was not lethal either.

In the third case, the injury must have been fatal: traces of a pointed (wooden) spear that struck the neck of the left femur and entered the pelvic cavity have been discovered among remains at Skhul (Mount Carmel in Israel) dating back 40,000 years. The injury was too serious for the victim to survive.

The First Massacre

The most ancient evidence of a real conflict (an actual massacre) is much more recent: it dates back to the Neolithic period. At Roaix in southeast France, what might be defined as a prehistoric paupers' grave was found. The bodies of men, women, and children of various ages were piled randomly on top of each other in an underground chamber. The homogeneity of the layer containing them and the perfect anatomical connection of the skeletons prove that the deaths were contemporaneous.

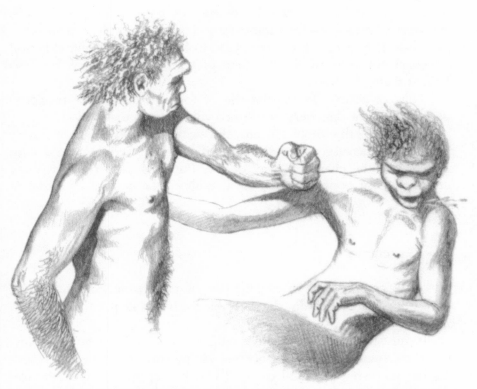

63. The fossil remains of the jawbone of a young boy who lived half a million years ago bear signs of a very violent blow to the right side of the face. Some people feel that this is the oldest evidence of human violence.

But examination of the skeletons provided the most dramatic information of all: they still had Neolithic arrowheads embedded in them. The massacre probably took place during a war, vendetta, or raid of some kind.

Shepherding and farming started during the Neolithic period; the first villages were established and the sedentary life with the rearing of domestic animals became more widespread. In other words, "property"—the possession of land, livestock, and food reserves—was conceived. Before that time, the nomadic life of our distant ancestors and the difficulty of transporting personal belongings (26 to 33 pounds of baggage at the most, as is the custom today among the Kung) had prevented or at least mitigated the sense of possession. Cows, fields, or granaries mean food for tomorrow and require less effort than hunting. Thus, it can lead to the temptation of theft. The Neolithic period witnessed not only the development of a sense of property, but also the first thefts and raids.

This, of course, led to the use of force in defense of property, as attested by the discovery of fortified villages from that period.

From Neolithic times onward, the signs of violence against people and objects increase progressively, as if violence were a natural consequence of the change in way of life. But it is also true that there are more fossil bones available as a result of the custom of burial of the dead.

Life Expectancy

While no certain data on the lifespan of prehistoric hominids exist, fossil bones and especially the wear on teeth can provide information on age.

It must be remembered that the life expectancy of an *erectus*, a *habilis*, or an *Australopithecus* was not necessarily similar to ours: each species has its own "biological clock" (dogs live up to twenty years, elephants over eighty, parrots more than a century, mice two to three years, cats just over ten, adult mayflies only one day). Thus, there are very marked genetic differences. There are also substantial differences among the primates: macaques live up to twenty-five years, gibbons up to thirty, chimpanzees over forty.

Our ancestors may have had different "biological clocks" (besides having a shorter lifespan because of the lack of medical care, hygiene, good nutrition, security, etc.). The average lifespan of Australopithecines has been calculated from fossil bones to be twenty to twenty-two years (Lucy was around twenty years old when she died); the *habilis* lived a little longer; the *erectus* from twenty to thirty years; and the Neanderthals up to forty or fifty years.

This does not mean that some hominids did not live longer (in Arago, for example, at a site 350,000 years old, some remains have been found of the mandible of a female *erectus* who apparently lived to be fifty years old); it does, however, underline the fact that natural selection was very high and that it was not easy to live to an old age in those conditions.

One interesting thing about the Neanderthal (for whom it is easier to calculate averages because of the great number of bones found) is that almost half the remains are of children. The many remains of eight- and nine-year-olds points to strong selection from childhood diseases.

But then again, it should not be forgotten that selection in the human species was harsh until very recently: the life expectancy in Italy just a century ago for those over five years of age was forty-three years.

Research carried out on the teeth of Australopithecines has also revealed a high rate of infant mortality, with death occurring particularly frequently between two and three years of age, that is, during weaning, the transition from the mother's breast to solid food. The researchers feel that the subsequent development of social behavior (and, therefore, collective nutritional efficiency provided in part by hunting) made this transition easier for children.

Finally, one factor must be kept in mind and that is, as mentioned at the beginning of this book, that fossilization is a very rare event. It only takes place when corpses are immediately covered with fine sediment (a typical example is drowning). Therefore, it is reasonable to wonder whether this has not created some kind of distortion in our conception: Are young people more prone to this kind of death? It is difficult to assess this factor, but it cannot be ignored.

Sometimes life in the wilderness, far from the pollution and poisons of industrial society, is thought to be healthy, clean, and serene. But just try to live in a forest or on the savannah for a year without weapons, antibiotics, a mosquito net, or tinned food, and to provide for your own food, water, and shelter each day. You'll soon see that it leads to premature aging (and death).

Appendix 3

What Teeth Can Tell under the Microscope

Prehistoric "Black Boxes"

Someone once said that "teeth are the first to decay when we're alive and the last when we're dead." Thanks to their high concentration of mineral salts (from 70 percent to 95 percent of dry weight, as compared to 65 percent in bones), teeth last much longer in sediments than bones. Their resistance and compactness often make them the only part of the bodies of our ancestors that remain intact, even after millions of years underground. They are like "black boxes" that have survived time, on which some of the daily habits of our ancestors, such as the food they normally ate, the way they chewed, the cleanliness of their teeth, the way they tanned skins, and so on, are recorded.

In trying to decodify the microscopic scratches, furrows, and wear on the surface of teeth, Sherlock Holmes the paleontologist now uses a number of very sophisticated techniques ranging from the electron microscope, able to make an imperceptible indenture look like a sprawling valley (which can then be studied like a geographic relief) to fine resins making extremely precise molds possible.

Many researchers have also studied the eating habits of other primates to see what and how they eat, and have compared the surface of their teeth with those of hominids. This has led to some very interesting conclusions about the lives of our ancestors.

As in all good detective stories, the culprit (food) has left some indirect evidence. It is well known, for example, that the crushing action of the molars inevitably causes the chipping of the surface of the

305

teeth, while tearing (generally done with the canines and incisors) leaves furrows and cuts.

These "fingerprints" left by food can be caused by abrasive particles contained in the food itself (along with compounds containing cellulose, liquid collagen, and some extremely long sugar chains—all extremely resistant—or the silica found in the cells of some plants) or grains of sand adhering to it (tubers, roots, etc.).

In this way it was discovered that the teeth of *afarensis* (that is, Lucy) show markings similar to those found on the teeth of chimpanzees (which are fond of fruit), and of gorillas (which often peel tough plants with their teeth) and show wear similar to that of the teeth of baboons (which eat a lot of seeds and small tough plant matter, causing the rapid exposure of the dentine).

In other words, microscopic analysis of the teeth of Lucy and of those like her has confirmed the hypotheses made during study of her bones and the deposits containing them: these beings were omnivores with vegetarian preferences, and their teeth were capable of chewing large quantities of plant matter, but not much meat.

In fact, the large incisors of *afarensis* were probably used to bite into and peel tough vegetables. And the prognathism they produce gave these ancient hominids an even more ape-like appearance.

Microscopic analysis of teeth has also revealed the later changes in diet that occurred as a consequence of the development of hunting. These are particularly evident in the teeth of *erectus*. When meat-eating becomes a habit, the teeth bear signs of it. For example, analysis of the cuts and furrows on the molars in half a mandible of an *erectus* found in the Tautavel Cave in France shows a remarkable similarity to those found on the teeth in some primitive populations, such as the Charruas in Uruguay, who are excellent hunters.

The First Toothpicks

The electron microscope has even made it possible to notice some strange markings on the teeth of some hominids that seem to indicate the use of toothpicks.

Numerous teeth of *erectus* found in a cave at Atapuerca, Spain, have very evident grooves between the crown and the roots. Microscopic analysis has revealed that they are V-shaped, as if they were formed by a long, narrow instrument passed back and forth across the teeth.

These findings are significant in that they attest to the use of rudimental toothpicks 400,000 years ago, probably to free the teeth of the bothersome bits of tendon left between them after eating meat. Then again, chimpanzees in the African forests have been observed to clean the teeth of their companions. A famous case was a chimpanzee that carefully polished the teeth of another (without poking into the spaces between them) with a piece of soft wood.

This behavior is a part of "grooming," the reciprocal cleaning activity that can often be observed among apes and which has considerable social importance in strengthening ties and emphasizing subjugation.

There is no reason to rule out that this habit of cleaning teeth could have been present among much earlier forms of hominids, but only took on particular importance when meat-eating became the custom. Someone has even suggested that the disastrous conditions of the teeth of Broken Hill Man (who lived in Zambia approximately 130,000 years ago)—devoured by caries and abscesses affecting the bone—were caused by his personal inability to keep his teeth clean.

Fermentation and suppuration were (and still are) the worst enemies of healthy teeth: that is why dental hygiene, achieved through use of the toothbrush and dental floss, is considered the best way to keep teeth in good shape.

It should, however, be pointed out that caries was not frequent among prehistoric beings (we will see why later). Their major problem was loss of teeth.

Losing Teeth in Ancient Times

Loss of teeth was mainly the result of the impoverishment of the surrounding bone. Diet shortages of some substances such as vitamin C (which is fundamental for the blood, but also for the bones) made individuals vulnerable: their teeth would become loose and then painful abscesses would form, making it difficult to chew. The most effective cure was extraction. It is quite likely that extraction was already practiced during prehistoric times, but we have no proof (even though in certain finds, such as those at Tautavel, the teeth are healthy but some are missing).

Missing teeth could cause serious problems: some skulls show that the individuals (mostly Neanderthals) were left with only a few teeth. One immediately wonders how they survived, especially since some lived to be quite old (over forty).

Perhaps some "tender" foods, such as soups and boiled meat, were already known at the time. Or perhaps food was prechewed by others. In this regard, the observation by ethnologist Eibl Eibesfeldt of the Max Planck Institute is interesting: some mothers prechew food for their children today: in the absence of teeth and baby food, the only available pap after weaning is supplied by the mother (Eibesfeldt considers this mouth-to-mouth feeding to be the origin of kissing).

In any case, prehistoric beings had an extraordinary defense mechanism which has been lost today. It can be observed under the microscope on some fossil teeth. In response to wear or caries, the teeth produced a protective coating. Dr. Pierre-François Puech, an expert in hominid teeth, explains:

> Our teeth are no longer able to react to adversity. In prehistoric times, when the teeth of a human being or an animal wore down, "apposition" took place, that is, a layer of calcified tissue like dentine formed as a reaction. This occurred in the canal, with the nerve retracting progressively as the tooth was consumed.

This self-defense mechanism must have alleviated the pain caused by tooth erosion, which was very severe at the time (although the short life expectancy reduced the chances of the teeth wearing down completely).

Smaller with More Caries

We will not go into the results of the other studies on teeth carried out through paleontological research and discussed elsewhere in this book. We would simply like to add two curious observations about the evolution of teeth.

The first is that looking at the "genealogical tree" of teeth, it can be seen that the decline of the third molar, the so-called wisdom tooth, started in very ancient times—with *Homo habilis*, about two million years ago.

The second concerns the rate at which teeth have become smaller in the last 100,000 years. Studies carried out by Loring Brace show that the surface area of incisors has decreased from 144 to 80 square millimeters and that of the third molar (the wisdom tooth) from 260 to 200 square millimeters.

This decrease has not been gradual: in the first 90,000 years it was 1 percent every 2,000 years. In the last 10,000 years, the rate of decrease has doubled. The preparation of foods, cooking in vessels, and, finally, the emergence of agriculture reduced the need for strong teeth. The millstones in the mouth were replaced by external mills in stone, and the strength and size of teeth were no longer criteria in selection.

This change in diet resulting from the development of agriculture was accompanied by an enormous drawback: the rapid increase in caries. In prehistoric times, caries was very rare. But from Neolithic times —approximately 7,000 to 10,000 years ago—with the introduction of more sugar into the diet (through the use of cereals and their fermentation), caries spread rapidly. This was the price to be paid for a more energy-rich diet.

Appendix 4

The Advent of Tools

A Parallel Story

"The shape of tools reflects the shape of the brain." In a certain sense this remark is true. If the various tools used during prehistory and the brains that evolved during that period were traced on two parallel lines, their synchronous evolution would immediately be evident.

Simple brains, like those of *Homo habilis*, led to the creation of simple instruments like "choppers." More developed brains, like those of *erectus*, allowed for the invention of increasingly sophisticated hand axes. The Neanderthals refined their manufacture even more, and the Cro-Magnon, who had a brain like that of human beings today, made by far the most elaborate and efficient tools.

Fig. 37 shows the growing complexity in the various stages of working stone: the increase over time in the number of operations carried out in manufacturing a tool can be seen, as can the number of blows for each operation.

From a certain point onward, cultural evolution affects the efficiency of each movement: modern-day human beings (with brains similar to those of the Cro-Magnons) can, for example, build a computer with a relatively small number of "blows"; but that is only apparently the case because behind it are 20 to 30 years of mental "pounding" to learn the know-how accumulated over centuries of study and invention.

In the following pages, we will try to reconstruct the story of the origin of tools, which is basically the origin of technology. We will

311

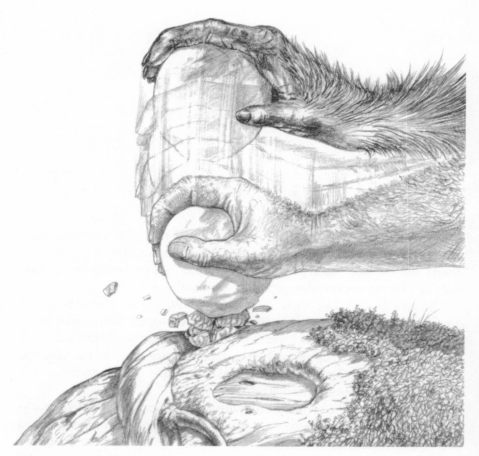

64. Using a rock to crack a nut is a familiar gesture. We know how to do that. The chimpanzees do, too. Most likely the first hominids also used natural tools before starting to work them.

not enter into the details of each working process with a pedantic (albeit important) list of classifications. What we will try to do is bring the history of tools to life in an overview and an examination of a few "magic" moments—those in which the shape of the object reflects a new shape of the brain.

The starting point is not the first chopper: the first chopper is already the end point of a process that started much earlier, at a time when hominids used unworked tools. Then again, human beings were not the first creatures to use tools; many animals preceded them. What we see in nature today confirms this.

Let's start out with a brief "glimpse." The main figure in the story will soon become evident.

A Blow in Time

A hairy hand with dark fingernails hovers over a cola nut no more than a few centimeters in diameter. The dark eyes, peering out of the fair-skinned and wrinkled face, study it.

Then the hand moves. First almost gropingly, then with increasingly coordinated movements, the nut is picked up. The fingers close around the nut and raise it to the eyes for closer examination. The head bends to one side. Then the crouching figure moves over to an exposed root of the nut tree.

The hairy hand delicately places the nut into a cavity already used for this purpose. The other hand grasps a large oval rock and raises it. A few seconds later, the sound of the sharp blow crushing the nut rings through the forest. For a moment nothing happens. Then the fingers delicately gather the edible pieces left of the nut.

This scene could have taken place 3 or 4 million years ago. But actually it's contemporary and the main figure is our cousin the chimpanzee.

Episodes like this are very common and have been described by numerous researchers such as Sugiyama, Koman, or Christophe, and Hedwige Boesh who have long studied chimpanzees. Perhaps the most important fact is that if the chimp wants to break the nut, it has to carry the rock with it, as there is none nearby.

Chimpanzees have been known to gather as many as a dozen cola nuts and then carry them to a hard surface (a rock or a root) to break them open in succession. Either way, the chimp does not act instinctively: a plan and, therefore, reasoning is required. Furthermore, it may have to carry the nuts and the stone hundreds of feet in order to implement its plan.

This behavior may be surprising, but it is not unique. Jane Goodall, the famous primatologist who studied chimpanzees in the Gombe

National Park in Tanzania for years, made the famous discovery that chimpanzees fish directly for the termites they are so fond of in the tunnels of termite hills by introducing branches stripped of leaves or blades of grass, waiting for the termites to attach to them and then pulling them out and licking them. The whole sequence is similar to an oil check in a car, only that the tortuousness of the tunnels in a termite hill makes it more difficult. Of course, this says a lot about animals that have a brain volume that is only a third of ours, and that have much less manual dexterity. In fact, chimpanzees are known to throw sticks, branches, and rocks to frighten adversaries or intruders and chew leaves to make sponges used to extract liquids from inaccessible places like holes in rocks or hollow tree trunks.

At first glance, these actions may seem very "human" for an ape, but many other animals are also capable of using "tools" in a surprising manner.

Sea otters are a well-known example: they use rocks gathered from the sea bottom to break open bivalves for their meals. Egyptian vultures, on the other hand, break open eggs with stones held in their beak. Crows do the same to break open nuts. And the well-known finches that Darwin observed on the Galapagos Islands use long thorns to dig around under the bark of trees and "harpoon" the larvae and worms they feed on.

If chimpanzees use simple tools intelligently today, it is logical to assume that our direct ancestors were able to do the same.

In other words, the use of tools did not arise suddenly—there was no "zero hour" in which their use began. There was probably a slow transition toward greater use of a pre-existing behavior. With the hominids, this behavior suddenly flourished. To use the technical term already used many times in this book, it can be said that there was a "pre-adaptation." Going back to the poker analogy, it is as though a player were dealt an excellent hand: he has an advantage with respect to the others, provided he knows how to put it to use.

It is difficult to establish who was the first to brandish a stick or a stone. And at this point, it is rather irrelevant. It was probably some anthropomorphus ape, before the genealogical line of the chimpanzee and our line separated; exactly how much earlier is impossible to say.

Wood and Bone

Before the manufacture of tools, our most ancient ancestors probably used "natural" tools such as branches and stones for their everyday needs.

Wood is obviously much easier to work than stone and its form immediately suggests a number of uses ("long" to reach inaccessible places, "massive" for beating, "pointed" for digging, etc.) Unfortunately, wood very rarely fossilizes. Thus, it is very difficult to find wooden tools.

The most ancient is a lancehead made of yew, found at Clacton-on-Sea in Sussex, England, that dates back 230,000 years (some say 400,000). Another ancient and dramatic find is a seven-foot yew lance found embedded in the ribcage of an elephant killed between 120,000 and 70,000 years ago in Lehringen, Germany. Accurate examination of the "murder weapon" has revealed that its point was brought to lethal hardness by fire. The only complete wooden tools that have been found date from 10,000 to 12,000 years ago.

Then, of course, there's bone. In the famous movie *2001: A Space Odyssey*, the first scenes show a hominid (perhaps an Australopithecine) picking up a bone and wielding it like a club.

We do not know how and when bones were first used as tools, but these scenes bring to mind a very important fact: our ancestors had an extraordinary range of implements to choose from—the bones of the animals on the savannah. It doesn't take much to come upon the idea of using a femur as a club, a row of teeth as a saw, or a bone chip or a horn as a dagger.

According to C. K. Brain and R. Susman, the most ancient evidence of this kind of use of bone can be found in the cave in Swartkrans in South Africa. Twenty-five or thirty bone splinters seem to be worn down at one end. Under the microscope, the scratches and abrasions are strikingly similar to those produced experimentally on bone by digging (this was discussed in the chapter on Australopithecines).

That means that almost two million years ago, someone used those bones to dig—probably for tubers—near the cave. Who were they? *Homo habilis? Australopithecus robustus?* We will never know. But we do know that *erectus* later used bones as tools.

We have evidence, for example, in Italy. Approximately 300,000 years ago, in the vicinity of what was later to become Rome, some early beings splintered the bones of elephants, deer, giant oxen, and other animals for later use. The knives, scrapers, graters, and even hand axes made of bone found at Castel di Guido and Malagrotta constitute an almost unique find in the history of human paleontology.

Why the preference for bone? Simply because there was no stone suitable for making tools in the area and bone was the best material available.

Of course, stone was used most often from the beginning (or to be more exact, it has been preserved the best). Today, paleontologists have recovered an enormous quantity of stone tools, which provide fascinating information about the past.

To what extent do these artifacts allow us to reconstruct the behaviors, the habits, the lifestyles, and the intelligence of our distant ancestors?

A Fossil Encyclopedia

Let's put the question a different way. Suppose that in a very distant future some archeologists from another planet were to land on the moon and find a camera forgotten by two American astronauts. What would the camera tell them about the past?

Lots of things. Apart from the film (which would probably have been ruined by cosmic rays), the pieces of the camera would provide enormous amounts of information. For example, analysis of the metal alloys or the structure of the lenses would give insight into the makers' state of knowledge at the time of the mission, their technological ability, and even their esthetic tastes.

That camera would become a tiny encyclopedia allowing the researchers of the future to return to our world. The same thing would happen if future extraterrestrial civilizations were to intercept space probes like the *Voyager* or the *Pioneer*, which carry a load of images, equipment, and technology on a route leading out of the solar system.

In other words, throughout the history of mankind, every civilization has left "messages" (encyclopedias) of its culture in the form of technology, works of art, and monuments. Or more simply in the form of stone tools that are low in technology, but rich in information.

Stone tools provide a precious source of information for study of our origins. For one thing, a stone tool is concrete proof of the presence of human beings in a certain area or a given geological stratum. It's like finding a pair of eyeglasses on a desert island.

It took almost thirty years to find the first hominid at Olduvai, but hundreds of tools found in the sediments attested to their presence.

But above all, stone tools are a kind of projection of the brain. It is like reading the composition of a student and trying to figure out whether he/she attends elementary school, secondary school, or university, or is illiterate.

One common error has to be avoided: primitive tools were not necessarily manufactured by people lacking intelligence. As Henri de Lumley points out, "Basically, if we have to cut a rope or open a tin on the beach and we don't have the right utensils, we use a sharp stone, like our ancestors did two million years ago."

There are still populations today that chip stone the same way *Homo erectus* did, although they have brains that are much more evolved. It must also be added that the primitiveness of an implement can depend on material that is difficult to work. But sometimes there are no doubts about the cultural level indicated by the tool. This is the case with the more ancient tools. Exactly how far back do the first stone tools go?

The Origin of Technology

We have to go to the desolate Afar Triangle in Ethiopia to answer that question. Here, in the scorching heat of the site at Gona, chips of simple basalt, dating back 2.3–2.4 million years (perhaps even 2.5 million years) have been found. Numerous fragments of quartz, which seem to be the product of ancestors from the same period, have also been found in the Omo River valley.

These chips and fragments are so rough that no lay person would take notice of them. The small hominids of the savannah chipped stones and pebbles without realizing that they were starting a process that is unique in nature and typical of the human race: technology, that is, the use of one tool to construct another. No animal knows how to do that.

Unfortunately we don't know who these first hominids were because no hominid bones have been found along with the tools. It is as though we have found the cigarette butt and don't know who smoked the cigarette.

We do have two suspects—*Australopithecus afarensis* and *Homo habilis*—but both have good alibis. The former lived half a million years prior to the most ancient tools and the former half a million years later. Who, then, gave rise to technology? A more evolved *Australopithecus* that has not yet been discovered? Or an archaic *habilis* that is also still hidden in the sediments? Or, as many claim, a transition form between the two?

We have no answer here, either. But we can attempt some conjectures on how the use of stone tools began. Let's start with the hairy hand wielding a rock to break a nut described at the beginning of this appendix. Even our most distant ancestors surely knew how to do whatever a chimpanzee can do.

It is quite likely that everything started in a very accidental way. It's a little far-fetched to think that one fine morning someone decided to chip a stone and use it as a tool. It is much more likely that our ancestors initially used chips or stones found on the ground or created accidentally while they tried to crack a nut on a rock.

Once they noticed the efficiency of the tool, they would have preferred it and looked for others like it. They may even have tried to refashion it by breaking other stones. This probably took place over a long period of time; a first technique may have been invented and then forgotten by one group and then reinvented by another and so on until it became a consolidated tradition.

At some point, however, around two million years ago (perhaps a little earlier), a lot of these sharp, chipped stones appeared—well-designed although still rudimental.

From that moment onward, simple stones were no longer enough; cutting surfaces were needed. This led to the appearance of the so-called choppers and chopping tools in Africa. These are stones that have been hewn on one side (chopper) or both sides (chopping tool) to make a cutting edge. These extremely rough tools vary in size from that of a mandarin to that of a large orange, meaning they fit nicely in the palm of the hand (or both hands).

Thus, it is not difficult to imagine what they were used for: cutting, crushing, or breaking wood and, above all, bone.

How do we know that? Well, the regular use of choppers and chopping tools coincides with the emergence of the most ancient representative or our genus *Homo*: *Homo habilis*, about two million years ago. These beings had the intelligence not only to take this first technological step but also to change diet (as is proven by study of their teeth)—from opportunistic vegetarians to omnivores. They probably took advantage of the spoils of lions, breaking open the bones with their tools to get at the nourishing marrow. These tools were also used daily to dig and grind tubers, to crush seeds, and so on.

It is surprising to find out how sharp a stone cutting edge can be: modern reproductions demonstrate that it cuts better than a steel blade (but has to be honed more frequently).

This stage of culture and technology has been called "Oldowan," after the site by the same name in the Olduvai Gorge where the first two million-year-old choppers and chopping tools were found.

Stone Drops

But there is another extremely interesting aspect to the birth of stone technology.

If we were to fashion a stone tool—a simple chopper, for example—we would notice that a large quantity of very sharp chips accumulates. We would instinctively think of using them as cutting tools, but what did our ancestors do? A sharp chip would be perfect for cutting skin, quartering meat, or severing ligaments.

Many experts feel that it is unlikely that our ancestors did not realize the usefulness of the chips. In fact, they claim that the chips from choppers were used as much as the chopper itself.

This led to a gradual improvement in the chipping technique. With the skillful and calibrated blows of *Homo habilis* and later *Homo erectus*, these implements were honed on both sides and lengthened into a very special shape—a drop or an almond. These tools are called "hand axes"; there is still much mystery about them today. What were they used for? How were they held?

As strange as it may seem, no answers to these questions have yet been found (some may have had wooden handles like those of Indian tomahawks.) Actually the label "hand axe" applies to a number

65. The evolution of tools.

of tools. Just as different kinds of knives are used to cut bread, meat, fish, or cheese, the larger and heavier hand axes probably served to crush bones and wood or to dig, while the smaller and sharper ones were used for cutting. Others may have been used for a combination of these purposes, like a Swiss army knife.

But some hand axes seem too beautiful and complex for simple daily use. Did they have special value? Were they some kind of prehistoric status symbol?

What is surprising is that the concept behind this tool remained unaltered for almost one and a half million years: the hallmark of *Homo erectus* all over the world.

The last important consideration that can be made about these stone tools is that they signal the beginnings of human imagination—that is, the ability to "see" something that does not yet exist, the ability to design an object that can only be created by a series of actions transforming a mental model into reality.

In the pebbles they picked up, the chunks of lava, limestone, flint, or quartzite they turned over in their hands, *Homo erectus* already "saw" the future tools, just as Michelangelo could see *David*, the *Pieta*, or *Moses* in the blocks of marble he examined.

Homo erectus also had their creations in mind. This ability to imagine, design, and fashion tools shows how far along the road of intelligence these toolmakers had come.

The Development of Techniques

The evolution of hand axes to bifacial tools to even more sophisticated implements was natural (although not rapid). The range of tools expanded: the chips were worked to make very sharp edges or special shapes that were used for scraping, grating, and cutting. Thus, tools called scrapers, grattoirs (end-scrapers), and points appeared. They made the cutting and cleaning of a carcass or the preparation of a wooden spear easier.

Researchers have named this cultural and technological period "Acheulian"; it extends from 100,000 to 1.4 million years ago, depending on the geographic areas.

When the *erectus* left Africa, they naturally took their toolmaking techniques with them (although there was some delay—the "cultural lag" referred to in the preceding chapter). But these techniques were eventually supplanted by a new technique (which appeared at the time of the last *Homo erectus*): the Levallois technique, representing another cultural step forward.

The name is derived from a suburb of Paris in which the first tools of this kind were found. Using this technique, a small piece of flint, basalt, or other kind of rock is prepared with a number of special blows. Then, with one final well-placed blow, a last flake comes off: this flake is the tool and its perfect shape is the direct result of the previous blows.

This technique is similar to those used by diamond cutters under special circumstances. It takes a lot of imagination and skill and can be used to obtain tools of different shapes (round, oval, rectangular, or pointed) from the same core (rather like a prehistoric assembly line).

This technique was adopted mainly by the last *erectus* (or archaic *sapiens*) and by the Neanderthals. The latter developed a characteristic industry (called "Mousterian") that strangely enough produced a large number of scrapers. These tools are often the size of a large timepiece or a pocket comb, that is, rather large, attesting to the lack of precision in Neanderthal craftsmanship and, above all, the power of their grip. But why so many scrapers? And why were these tools dominant for tens of thousands of years?

The Neanderthals were excellent hunters. A scraper can be used for a multitude of purposes, but basically it served to butcher animals. Furthermore, scrapers were indispensable for skinning prey and furs were needed to survive the glacial climates.

At one point, in Europe approximately 35,000 years ago, the monotony of the Neanderthal style was shaken by a new concept in toolmaking: that of the *Homo sapiens sapiens* populations (Cro-Magnon, Chancelade, Combe Capelle). Occupying Europe, they dethroned the Neanderthals who, just before their disappearance, made a final attempt to renew their techniques (as described in chapter 14). The assemblages of the *sapiens sapiens* included a number of different typologies which rapidly succeeded one another: large scrapers were abandoned in favor of absolutely original tools such as burins for chiseling, gravers, punches, and, above all, long and sharp blades. Handles came into widespread use.

This technical revolution was possible thanks to new and very ingenious techniques. Blades, for example, were produced in a manner similar to the "Levallois" method, producing scores of sharp blades, rather like taking an infinite number of peels off an apple.

The famous "laurel" and "willow leaves" also date back to that time. These long, thin, leaf-shaped flint tools are true Paleolithic masterpieces.

These results were often achieved using tricks such as the heating of flint in the fire to make it more fragile and shiny, or knocking on a block of flint before working it to find out whether it is compact or has been attacked by frost or other agents.

Learning from Our Ancestors

Many researchers have tried to work flint, studying the chips and fragments formed and even the way they lie on the ground. They have found that unless one knows the technique, it is not at all easy to chip a piece of basalt, quartzite, or flint. The stone often breaks in the wrong way; more commonly, the craftsman ends up with injured fingers.

Experimentation with and the study of tools have led to the understanding that our ancestors simply struck two rocks together at first. Then they started to use so-called soft objects (such as wood, bone, and horn) to strike, producing more precise chipping with straighter edges and smaller chips. Later, a technique was developed in which the flint was no longer struck; it was put under pressure so that the pieces that came away were of exactly the desired shape and size, producing blades and the "laurel leaves" mentioned previously.

Experimental studies have also proven that it does not take long to make an efficient stone tool: a good hand axe, like those used by *erectus*, can be made in two or three minutes. It only takes about a half minute to make a scraper. Thus, it is erroneous to think that our ancestors spent hours of their time manufacturing tools.

But experimentation has gone one step farther. Researchers have examined the tiny grooves, the microabrasions and the microchipping caused by use, and have compared them to those found on the authentic tools found in the excavations. In this way they have discovered how some of the tools were actually used. For example, in many cases it was possible to determine whether the tool had been used to cut meat or rasp wood, to chisel bone or to scrape skins. In some exceptional cases, it was possible to distinguish whether the tool had been used on leather or on fresh skin, bone, horn, tender plant matter, or hard wood.

Many things still have to be understood about the manufacture and use of tools by hominids. In the beginning we mentioned that the shape of their implements reflected the shape of their brains. We can now add that our ability to understand them will reflect the shape of our brains, that is, the knowledge and competence that we will manage to introduce into the structure of our brains.

Bibliography

Aiello, Leslie. *Discovering the Origins of Mankind*. London: Trevin Coppleston Books Limited, 1986.

Aiello, Leslie, and Christopher Dean. *Human Evolutionary Anatomy*. London: Academic Press Limited, 1990.

Ardkey, Robert. *African Genesis*. New York: Atheneum, 1961.

Blumenschine, Robert. "Characteristics of an Early Hominid Scavenging Niche." *Current Anthropology*, August–October 1987.

Bordes, F. *Leçons sur le paléolithique*. 3 volumes. Paris: Editions du CNRS, 1984.

Brain, C. K. "The Evolution of Main in Africa: Was It a Consequence of Caiozoic Cooling?" A. Du Toit Memorial Lecture No. 17, Geological Society of South Africa, 1981.

Brown, F., J. Harris, R. Leakey, and A. Walker. "Early Homo erectus Skeleton from West Turkana, Kenya." *Nature* 316, no. 29 (August 4–September 4, 1985).

Bunn, Henry, and Helen M. Kroll. "Systematic Butchery by Plio-Pleistocene Hominids at Olduvai Gorge, Tanzania." *Current Anthropology* (December 1986).

Campbell, Bernard G. *Humankind Emerging*. Glenview, Ill.: Scott, Foresman and Company, 1988.

Clark, J. D. *The Prehistory of Africa*. London: Thames and Hudson, 1970.

Cronin, J. E., N. T. Boaz, C. B. Stringer, and Y. Rak. "Tempo and Mode in Hominid Evolution." *Nature* 292 (July 9, 1981).

Dart, Raymond. *Adventures with the Missing Link*. Philadelphia: The Institute Press, 1967.

Delson, Eric. *Ancestor: The Hard Evidence.* New York: Alan R. Liss, Inc., 1984.

Delta, Willis. *The Hominid Gang: Behind the Scenes in the Search for Human Origins.* New York: Penguin Books, 1991.

De Neandertal à Cro-Magnon. Catalogue édité par l'Association pour la Promotion de la Recherche Archéologique en Ile-de-France Musée de Préhistorire. Ile-de-France, 1989.

Elderedge, N., and I. Tattersall. *The Myths of Human Evolution.* New York: Columbia University Press, 1982.

Fagan, Brian M. *The Great Journey: The People of Ancient America.* New York: Thames and Hudson, Ltd., 1987.

———. *The Journey from Eden.* London: Thames and Hudson, Ltd., 1990.

Goodall, Jane. *The Chimpanzees of Gombe.* Cambridge, Mass.: Harvard University Press, 1986.

Gribbin, John, and Jeremy Cherfas. *The Monkey Puzzle.* Milan: Arnoldo Mondadori Editore, 1984.

Gribbin, John and Mary. *Children of the Ice: Climate and Human Origins.* Oxford: Basil Blackwell, 1990.

———. "Hominidae." *Proceedings of the 2nd International Congress of Human Paleontology.* Turin, September 28–October 3, 1987. Milan: Jaca Book, 1989.

Homo. Viaggio alle origini della storia. Testimonianze e reperti per 4 milioni di anni. Milan: Cataloghi Marsilo, 1895.

Howell, Clark. *Early Man.* Rev. ed. New York: Time-Life Books, 1973.

Isaac, Glynn. "The Food Sharing Behavior of Protohominids." *Scientific American,* April 1976.

Johanson, Donald, and Maitland Edey. *Blueprints: Solving the Mystery of Evolution.* Boston/Toronto/London: Little, Brown and Company, 1989.

———. *Lucy: The Beginning of Humankind.* New York: Warner Books, 1981.

Johanson, Donald, et al. "New Partial Skeleton of Homo habilis from Olduvai Gorge, Tanzania." *Nature,* May 21, 1987.

Johanson, Donald, and James Shreeve. *Lucy's Child: The Discovery of Human Ancestor.* New York: William Morrow and Company, 1989.

Kennet, Oakley P. *Man the Tool-Maker.* Chicago: The University of Chicago Press, 1972.

Kimbel, W., and T. D. White. "A Reconstruction of the Adult Cranium of Australopithecus afarensis." *American Journal of Physical Anthropology*, February 1980.

Kimbel, W., T. D. White, and D. Johanson. "Implication of KNMWT 17000 for the Evolution of 'Robust' Australopithecus." In *The Evolutionary History of the "Robust" Australopithecines*, F. E. Grine, ed., 1990.

Kinzey, Warren G. *The Evolution of Human Behavior: Primate Models.* State University of New York Press, 1987.

Leakey, L. S. B. "Exploring 1,750,000 Years into Man's Past." *National Geographic*, October 1961.

Leakey, L. S. B., P. V. Tobias, and J. R. Napier. "A New Species of the Genus Homo from Olduvai Gorge." *Nature*, April 1964.

Leakey, M. D. *Olduvai Gorge.* Volume 3. Cambridge, England: Cambridge University Press, 1972.

Leakey, M. D., et al. "Fossil Hominids from the Laetoli Beds." *Nature*, August 1976.

Leakey, R. E. F, and A. Walker. "On the Status of Australopithecus afarensis." *Science*, March 1980.

Leakey, Richard E., and Roger Lewin. *Origins: What New Discoveries Reveal about the Emergence of Our Species and Its Possible Future.* Roma-Bari: Laterza & Figli, 1979.

———. *The Making of Mankind.* London: The Rainbird Publishing Group Limited, 1982.

———. *People of the Lake: Mankind and its Beginnings.* New York: Avon Books, 1979.

Lewin, Roger. *Bones of Contention.* New York: Simon and Schuster, 1987.

———. "The Origin of the Human Mind." *Science*, May 8, 1987.

Lovejoy, C. Owen. "Evolution of Human Walking." *Scientific American*, November 1988.

Lumsden, Charles J. *Promethean Fire.* President and Fellows of Harvard College, 1983. Milan: Arnoldo Mondadori Editore, 1984.

Mellars, P., and C. Stringer. *The Human Revolution.* Princeton, N.J.: Princeton University Press, 1989.

Nelson, Harry, and Robert Jurmain. *Introduction to Physical Anthropology.* St. Paul, Minn.: West Publishing Company, 1988.

Pilbeam, D. "The Descent of Hominoids and Hominids." *Scientific American* 250 (March 1984).

Pilbeam, D. "Pilocene Hominid Fossils from Hadar, Ethiopia." *American Journal of Physical Anthropology* 57, no. 4 (April 1982).

Potts, Richard, and Pat Shipman. "Cutmarks Made by Stone Tools on Bones from the Olduvai Gorge, Tanzania." *Nature*, June 18, 1981.

Rak, Yoel. "Australopithecine Taxonomy and Phylogeny in Light of Facial Morphology." *American Journal of Physical Anthropology* (1985).

Reader, John. *Missing Links*. William Collins Sons & Co., 1983.

Renfrew, Colin. *Before Civilization: The Radiocarbon Revolution and Prehistoric Europe*. Cambridge University Press, 1979.

Sagan, Carl, and Ann Druyan. "Can Chimps Use Language?" *Sky*, October 1992, pp. 76–85.

Semah, François and Anne Marie, and Tony Djubiantono. *They Discovered Java*. Jakarta: Pusat Penelitan Arkologi Nasional, 1990.

Sergi, Sergio. *Il cranio neandertaliano del Monte Circeo*. Rome: Accademia Nazionale del Lincei, 1974.

Sieveking, A. *The Cave Artists*. London: Thames and Hudson, 1979.

Simons, E. L., *Primate Evolution: An Introduction to Man's Place in Nature*. London and New York: Macmillan, 1972.

Stringer, C. B. "The Credibility of Homo habilis." In *Major Topics in Primate and Human Evolution*, B. Wood, L. Martin, and P. Andrews, eds. Cambridge University Press, 1984.

Szalay, F. S., and Eric Delson. *Evolutionary History of the Primates*. New York and London: Academic Press, 1979.

Tianyuan, Li, and Dennis A. Eller. "New Middle Pleistocene Hominid Crania from Yunxian in China." *Nature* 357 (1992): 404–407.

Vrba, Elisabeth. *The Environmental Context of the Evolution of Early Hominids and their Culture*. In press.

Walker A., R. E. Leakey, J. M. Harris, and F. H. Brown. "2.5 Myr Australopithecus boisei from West of Lake Turkana, Kenya." *Nature*, August 7, 1986.

Weiss, Mark L., and Alan E. Mann. *Human Biology and Behavior: An Anthropological Perspective*. Boston: Little, Brown & Company, Limited, 1981.

White, Tim D., and Gen Suwa. "Hominid Footprints at Laetoli: Facts and Interpretations." *American Journal of Physical Anthropology*, April 1987.

Wymer, John. *The Palaeolithic Age*. New York: St. Martin's Press, 1982.